AF882473

Experimental Nephrology

Journal of Renal Cell Biology

KARGER

Printed in Switzerland on acid-free paper by Reinhardt Druck, Basel

Appears bimonthly: 1 volume per year (6 issues)

Guidelines for Authors

Experimental Nephrology

Submission

Only original papers written in English are considered and should be sent to:

Prof. L.G. Fine
Royal Free and University College
Medical School
5th Floor, Jules Thorn Institute
The Middlesex Hospital
Mortimer Street
London W1N 8AA (UK)
Fax +44 20 7380 9117
E-Mail j.deluca@ucl.ac.uk

Manuscripts should be submitted in quadruplicate (with four sets of original illustrations), typewritten double-spaced on one side of the paper, with a wide margin.

Conditions

All manuscripts are subject to editorial review. Manuscripts are received with the explicit understanding that they are not under simultaneous consideration by any other publication. Submission of an article for publication implies the transfer of the copyright from the author to the publisher upon acceptance. Accepted papers become the permanent property of 'Experimental Nephrology' and may not be reproduced by any means, in whole or in part, without the written consent of the publisher. It is the author's responsibility to obtain permission to reproduce illustrations, tables, etc. from other publications.

Arrangement

Title page: The first page of each paper should indicate the title, the authors' names, and the institute where the work was conducted. A short title for use as running head as well as the full address of the author to whom correspondence should be sent are also required.

Full address: The exact postal address complete with postal code must be given at the bottom of the title page. Please also supply phone and fax numbers, as well as your e-mail address.

Key words: For indexing purposes, a list of 3–10 key words in English is essential.

Abstract: Each paper needs an abstract of up to 200 words. It should be organized as follows:

Background/Aims:	What is the major problem that prompted the study?
Methods:	How was the study done?
Results:	Most important findings?
Conclusion:	Most important conclusion?

Small type: Paragraphs which can or must be set in smaller type (case histories, test methods, etc.) should be indicated with a 'p' (petit) in the margin on the left-hand side.

Footnotes: Avoid footnotes. When essential, they are numbered consecutively and typed at the foot of the appropriate page.

Tables and illustrations: Tables and illustrations (both numbered in Arabic numerals) should be prepared on separate sheets. Tables require a heading and figures a legend, also prepared on a separate sheet. For the reproduction of illustrations, only good drawings and original photographs can be accepted; negatives or photocopies cannot be used. Due to technical reasons, figures with a screen background should not be submitted. When possible, group several illustrations on one block for reproduction (max. size 181 × 223 mm) or provide crop marks. On the back of each illustration, indicate its number, the author's name, and 'top' with a soft pencil.

Color illustrations: Up to 6 color illustrations per page can be integrated within the text at the reduced price of CHF 660.– / USD 545.00 per page. Color illustrations are reproduced at the author's expense.

References: In the text identify references by Arabic numerals [in square brackets]. Material submitted for publication but not yet accepted should be noted as 'unpublished data' and not be included in the reference list. The list of references should include only those publications which are cited in the text. Do not alphabetize; number references in the order in which they are first mentioned in the text. The surnames of the authors followed by initials should be given. There should be no punctuation other than a comma to separate the authors. Cite all authors, 'et al' is not sufficient. Abbreviate journal names according to the Index Medicus system. (Also see International Committee of Medical Journal Editors: Uniform requirements for manuscripts submitted to biomedical journals. N Engl J Med 1997;336: 309–315.)

Examples

(a) Papers published in periodicals: Sun J, Koto H, Chung KF: Interaction of ozone and allergen challenges on bronchial responsiveness and inflammation in sensitised guinea pigs. Int Arch Allergy Immunol 1997;112:191–195.

(b) Monographs: Matthews DE, Farewell VT: Using and Understanding Medical Statistics, ed 3, revised. Basel, Karger, 1996.

(c) Edited books: Parren PWHI, Burton DR: Antibodies against HIV-1 from phage display libraries: Mapping of an immune response and progress towards antiviral immunotherapy; in Capra JD (ed): Antibody Engineering. Chem Immunol. Basel, Karger, 1997, vol 65, pp 18–56.

Galley proofs

Unless indicated otherwise, galley proofs are sent to the first-named author and should be returned with the least possible delay. Alterations made in galley proofs, other than the correction of printer's errors, are charged to the author. No page proofs are supplied.

Reprints

Order forms and a price list are sent with the galley proofs. Orders submitted after the issue is printed are subject to considerably higher prices.

KARGER

Fax + 41 61 306 12 34
E-Mail karger@karger.ch
www.karger.com

The Guidelines for Authors are available at:
www.karger.com/journals/exn/exn_gl.htm

Experimental
Nephrology

General Information

Publication data: 'Experimental Nephrology' is published 6 times a year. Volume 8 with 6 issues appears in 2000.
Subscription rates: Subscriptions run for a full calendar year. Prices are given per volume.
Personal subscription:
Subscription + postage and handling
CHF 289.10 + CHF 25.90 Europe
CHF 289.10 + CHF 39.10 Overseas
USD 220.85 + USD 25.90
+ online subscription per volume:
CHF 39.– / USD 30.00
(Must be in the name of, billed to, and paid by an individual. Order must be marked 'personal subscription'.)
Institutional subscription:
Subscription + postage and handling
CHF 826.– + CHF 32.30 Europe
CHF 826.– + CHF 48.80 Overseas
USD 631.00 + USD 32.30
+ online subscription per volume:
CHF 58.– / USD 45.00
Airmail postage: CHF 36.–, USD 30.00.
(Extra per volume.)
Discount subscription prices:
Members of nephrological societies.

Microforms: Please direct all enquiries and orders to: University Microfilms Inc.
300 North Zeeb Road, P.O. Box 1346
Ann Arbor, MI 48106-1346 (USA)
Subscription orders: Orders can be placed at agencies, bookstores, directly with the Publisher
S. Karger AG, P.O. Box,
CH–4009 Basel (Switzerland)
or with any of the following distributors:
Deutschland: S. Karger GmbH, Postfach
D–79095 Freiburg
France: Librairie Luginbühl
36, bd de Latour-Maubourg, F–75007 Paris
Great Britain, Ireland: S. Karger Publishers
58 Grove Hill Road
Tunbridge Wells, Kent TN1 1SP, UK
USA: S. Karger Publishers, Inc.
26 West Avon Road, P.O. Box 529
Farmington, CT 06085, USA
India, Bangladesh, Sri Lanka:
Panther Publishers Private Ltd.
39, 6th Cross
Wilson Gardens
Bangalore 560 027, India
Japan: Katakura Libri, Inc.
2F, Marushima Bldg, 4-6-17, Yushima,
Bunkyo-ku, Tokyo 113-0034, Japan

Change of address: Both old and new address should be sent to the subscription source.
Single issues and back volumes: Information can be obtained through the Publisher.
Advertising: Correspondence and rate requests should be addressed to the Publisher.
Bibliographic indices: This journal is regularly listed in bibliographic services, including *Current Contents®*.
Copying: This journal has been registered with the Copyright Clearance Center (CCC), as indicated by the code appearing on the first page of each article.
For readers in the US, this code signals consent for copying of articles for personal or internal use, or for the personal or internal use of specific clients, provided that the stated fee is paid per copy directly to
Copyright Clearance Center Inc.
222 Rosewood Drive
Danvers, MA 01923 (USA)
A copy of the first page of the article must accompany payment. Consent does not extend to copying for general distribution, for promotion, for creating new works, or for resale. In these cases, specific written permission must be obtained from the copyright owner,
S. Karger AG,
P.O. Box, CH–4009 Basel (Switzerland).

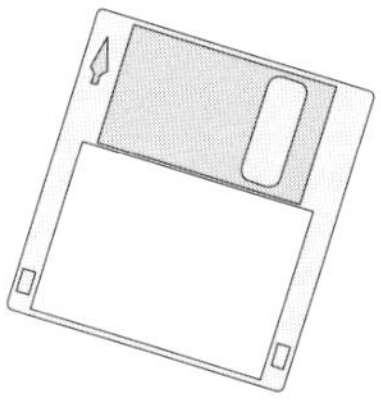

Disk Submission

A **comprehensive version** of these guidelines for the **submission of manuscripts and illustrations on disk** can be accessed online under **www.karger.com/journals/disksub.htm** or requested as a leaflet from the editorial department.

Please follow the general rules on style and arrangement in the **'Guidelines for Authors'** of the journal. Submit a **double-spaced printout** of your manuscript matching the file on disk exactly.

Technical Information
Store only data pertinent to this manuscript. Include your fonts (screen and printer fonts for Macintosh). For a **text only** manuscript, use a 3.5-inch disk. For **text and illustrations,** use 3.5 or 5.25 Magneto Optical Disc; CD-ROM (DOS or hybrid); iomage Zip/DOS (100 MB), or 44 MB, 88 MB or 200 MB SyQuest. Compress extensive data using ZIP format.

Data Description
Label your disk with the **title of the journal or book** for which it is submitted; a **short version** of the paper's title; your **name,** postal and e-mail address, fax and phone number; **hard and software** used, including the version; and the **complete list of the file names.** File names consist of eight characters before the dot identifying the file and three characters after the dot identifying the application.

Text and Format
The **preferred word-processing package is MS-Word.** Other commonly used PC text programs and ASCII formats are accepted.
Update your data when you revise subsequent to the reviewer's or editor's decision and mark revisions clearly. Manuscripts accepted for publication are subject to **copy editing:** changes and/or questions will be marked in the manuscript, which will be returned to you with your disk and the galley proofs for checking.
Enter your text continuously flush left. Do not use any layout functions. Do not split words at the end of a line. **Do not use indentations.** Use a blank line before and a hard return at the end of paragraphs.
Arrange headings flush left. Insert only **one space after punctuation marks.** Insert Greek letters, mathematical symbols etc. with your word-processing program. Use **boldface and italics** as well as sub- and superscript where appropriate.
Tables are part of the text and should be placed at the end of the text file. Separate the individual parts of a table by lines. Use the tabulator instead of spaces. Format operation data on one line in the same column. **Legends to illustrations are part of the text** and should also be listed at the end of the text file.

Illustrations
Store illustrations as separate files: do not integrate them in your text. Export **line drawings/vector graphics** as TIF, EPS or WMF format. Use Photoshop for processing and retouching **scanned halftone images.** Save the original scan and the processed version. Export **black and white or color images** as TIF or EPS format in their anticipated size in print.
For detailed information on how to submit illustration data on disk please consult the leaflet or **www.karger.com/journals/disksub.htm.**

KARGER
Fax +41 61 306 12 34
E-Mail karger@karger.ch
www.karger.com

The Journal Home Page is available at:
www.karger.com/journals/exn

KI 98 DISK

Experimental
Nephrology

Renal Ion Channels: Function and Regulation

Guest Editor
Rajesh V. Thakker, Oxford

19 figures, 4 tables, 2000

KARGER

Basel · Freiburg · Paris · London · New York ·
New Delhi · Bangkok · Singapore · Tokyo · Sydney

S. Karger
Medical and Scientific Publishers
Basel · Freiburg · Paris · London
New York · New Delhi · Bangkok
Singapore · Tokyo · Sydney

Drug Dosage
The authors and the publisher have exerted every effort to ensure that drug selection and dosage set forth in this text are in accord with current recommendations and practice at the time of publication. However, in view of ongoing research, changes in government regulations, and the constant flow of information relating to drug therapy and drug reactions, the reader is urged to check the package insert for each drug for any change in indications and dosage and for added warnings and precautions. This is particularly important when the recommended agent is a new and/or infrequently employed drug.

P.O. Box, CH–4009 Basel (Switzerland)
Printed in Switzerland on acid-free paper by
Reinhardt Druck, Basel
ISBN 3–8055–7143–7

KARGER

Fax+ 41 61 306 12 34
E-Mail karger@karger.ch
www.karger.com

Vol. 8, No. 6, 2000

Experimental Nephrology

Contents

319 **Preface**
Thakker, R.V. (Oxford)

320 **Disorders of the Epithelial Na^+ Channel in Liddle's Syndrome and Autosomal Recessive Pseudohypoaldosteronism Type 1**
Oh, Y.S.; Warnock, D.G. (Birmingham, Ala.)

326 **Defective Processing and Trafficking of Water Channels in Nephrogenic Diabetes insipidus**
Kamsteeg, E.J.; Deen, P.M.T.; van Os, C.H. (Nijmegen)

332 **Defects in Processing and Trafficking of Cystic Fibrosis Transmembrane Conductance Regulator**
Kunzelmann, K. (St. Lucia, Brisbane); Nitschke, R. (Freiburg)

343 **Mechanisms of Renal Calcium Transport**
Friedman, P.A. (Pittsburgh, Pa.)

351 **Molecular Pathology of Renal Chloride Channels in Dent's Disease and Bartter's Syndrome**
Thakker, R.V. (Oxford)

361 **Severely Impaired Urine-Concentrating Ability in Mice Lacking the CLC-K1 Chloride Channel**
Uchida, S.; Marumo, F. (Tokyo)

366 **Molecular Characteristics of Phosphate Transporters and Their Regulation**
Hernando, N.; Forster, I.C.; Biber, J.; Murer, H. (Zurich)

376 **Author Index Vol. 8, No. 6, 2000**
376 **Subject Index Vol. 8. No. 6, 2000**

377 **Author Index Vol. 8, 2000**
379 **Subject Index Vol. 8, 2000**
after 380 **Contents Vol. 8, 2000**

KARGER

Fax+ 41 61 306 12 34
E-Mail karger@karger.ch
www.karger.com

Access to full text and tables of contents, including tentative ones for forthcoming issues:
www.karger.com/journals/exn/exn_bk.htm

Experimental
Nephrology

Exp Nephrol 2000;8:319

Preface

In recent years there have been important advances in our knowledge of the mechanisms underlying the aetiology of many renal tubular disorders. These advances have resulted from the combined application of molecular biology and physiology, and the purpose of this special issue is to provide a succinct, readable overview of these advances. It would be impossible to be wholly comprehensive within one issue, so the aim has been to concentrate on some important developments. A series of authoritative accounts, written by international leaders in their fields have been gathered in this issue. These describe the advances that have been made in: understanding transepithelial sodium transport by analysis of the sodium channel (ENaC), and its role in Liddle's syndrome and autosomal recessive pseudohypoaldosteronism type 1; the manner in which aquaporin-2 transports water and how mutations result in autosomal recessive and dominant forms of nephrogenic diabetes insipidus; the essential role of the cystic fibrosis transmembrane conductance regulator (CFTR) in epithelial electrolyte transport and the resulting defects in processing and trafficking due to mutations; the novel insights in renal calcium transport gained from an analysis of channels in the proximal tubule, thick ascending limb of Henle's loop and the distal tubule; the identification of a new renal voltage-gated chloride channel (CLC-5) by positional cloning studies in Dent's disease and the role of CLC-Kb in a form of Bartter's syndrome; the role of the chloride channel, CLC-K1, in urinary concentration that was revealed by studies of a mouse knockout model; and the molecular characteristics of phosphate transporters and their roles in regulating this anion.

The reviews of these topics represent our current views of these areas and provide starting points for future continuing investigations. It has been a pleasure to work with my colleagues who have contributed so generously and enthusiastically of their own time and expertise in preparing and meeting the objectives of this issue.

R.V. Thakker

KARGER

Fax +41 61 306 12 34
E-Mail karger@karger.ch
www.karger.com

Accessible online at:
www.karger.com/journals/exn

Experimental Nephrology

Exp Nephrol 2000;8:320–325

Disorders of the Epithelial Na^+ Channel in Liddle's Syndrome and Autosomal Recessive Pseudohypoaldosteronism Type 1

Young S. Oh[a,b] David G. Warnock[a]

Departments of [a]Medicine and [b]Neurobiology, Nephrology Research and Training Center, University of Alabama at Birmingham, Ala., USA

Key Words

Epithelial Na^+ channel · Hypertension · Genetic defects · Pathophysiology

Abstract

The epithelial Na^+ channel (ENaC) is the key step in many Na^+-absorptive epithelia, such as kidney and distal colon, that controls the overall rate of transepithelial Na^+ transport. ENaC is composed of three homologous subunits, α, β, and γ. The α subunit is the key subunit for the formation of a functional ion channel, while the β and γ subunits can greatly potentiate the level of expressed Na^+ currents. ENaCs belong to the recently identified DEG/ENaC supergene family, sharing the same basic structure with cytoplasmic amino and carboxy termini, two transmembrane regions, and a large extracellular loop. The human ENaC genes have been cloned, and using genetic linkage analysis the involvement of ENaC gene mutations in two distinct human diseases, Liddle's syndrome and autosomal recessive pseudohypoaldosteronism type 1 (PHA-1), has been demonstrated. In Liddle's syndrome, gain-of-function mutations in the β or γ ENaC subunits have been found; all identified mutations so far reside in the carboxy terminus of the protein, either deleting or modifying the functionally important PY motif. In PHA-1, loss-of-function mutations in the α, β, or γ subunits have been found; these mutations either truncate a significant portion of the structure or modify an amino acid that plays an important role in channel function. In this review, our current understanding about ENaC and the pathophysiology of Liddle's syndrome and PHA-1 caused by ENaC mutations will be discussed.

This review is dedicated to 'GS', the originally described proband with Liddle's syndrome, who died on November 17, 1999, from the complications of chronic hepatitis C.

Epithelial Na^+ Channels and the DEG/ENaC Superfamily

The content of body fluid Na^+ is tightly regulated by the epithelial Na^+ channel (ENaC), which is present in the apical membranes of many Na^+-absorptive epithelia, such as the renal tubules, distal colon, and lungs. ENaC is specifically inhibited by the diuretic drug amiloride, and thus it is also called the amiloride-sensitive Na^+ channel [1, 2]. The major physiological role of ENaCs is to maintain Na^+ homeostasis, blood volume, and blood pressure by providing an apical entry pathway for Na^+ ions to permit rap-

KARGER

Fax + 41 61 306 12 34
E-Mail karger@karger.ch
www.karger.com

Accessible online at:
www.karger.com/journals/exn

Young Oh, PhD
Department of Medicine, Sparks Bldg 865, 1720 7th Ave. South
University of Alabama at Birmingham
Birmingham, AL 35294 (USA)
Tel. +1 205 934 3806, Fax +1 205 934 1147, E-Mail yoh@uab.edu

id vectorial transport of Na^+ from the luminal to the interstitial compartment in Na^+-absorptive epithelia, such as the kidney distal tubule segments (fig. 1). Three subunits of ENaC have been cloned, α, β, and γ, which are homologous to each other, having about 35% amino acid identity over the entire length of coding regions [3].

ENaCs share significant sequence similarity with the nematode *Caenorhabditis elegans* degenerin (DEG) genes, such as *mec-4*, *mec-10*, and *deg-1*, which are involved in sensory touch transduction and, when mutated, neuronal degeneration [4]. Therefore, it has been proposed that ENaC subunits and the *mec-4*, *mec-10*, and *deg-1* proteins all belong to a new gene family (DEG/ENaC superfamily) that encodes cation channels involved in the control of the cellular and extracellular volume [5]. Other members of the DEG/ENaC superfamily include the ENaC δ subunit cloned from the rat and human brain [6], mammalian degenerin or brain Na^+ channels [7], the acid-sensing ionic channel [8], and a FRMFamide-activated Na^+ channel cloned from the snail *Helix aspersa* nervous system [9]. The DEG/ENaC superfamily contains similar sequences and the same predicted structure: intracellular N- and C-termini, two hydrophobic membrane-spanning regions, and a large extracellular loop. This topology has been experimentally demonstrated for the ENaC α subunit [10–12] and MEC-4 [4, 13].

Functional Studies of ENaC

Functional expression studies have demonstrated that the three ENaC subunits form a heteromultimeric, constitutively active channel when coexpressed in *Xenopus* oocytes as well as in mammalian cells. The stoichiometry of the functional ENaC complex is not yet established, and it has been suggested to be either a tetramer [14, 15] or nanomer [16, 17]. The αENaC alone can produce an amiloride-sensitive Na^+ current in *Xenopus* oocytes, but the amplitude of the Na^+ current is small [3, 18]. Nevertheless, the critical role of αENaC in forming a functional Na^+ channel in vivo was demonstrated by knocking out the αENaC gene in mice that died within 40 h of birth due to the failure of lung liquid clearance [19]; it is known that ENaC plays a critical role in clearing lung liquid during birth [20]. The role of αENaC for proper subunit assembly and targeting of the assembled channel complex to the apical membrane of epithelial cells has been described [21–23]. Several naturally occurring splice variants of human αENaC have been identified [24, 25], but their physiological significance remains to be characterized.

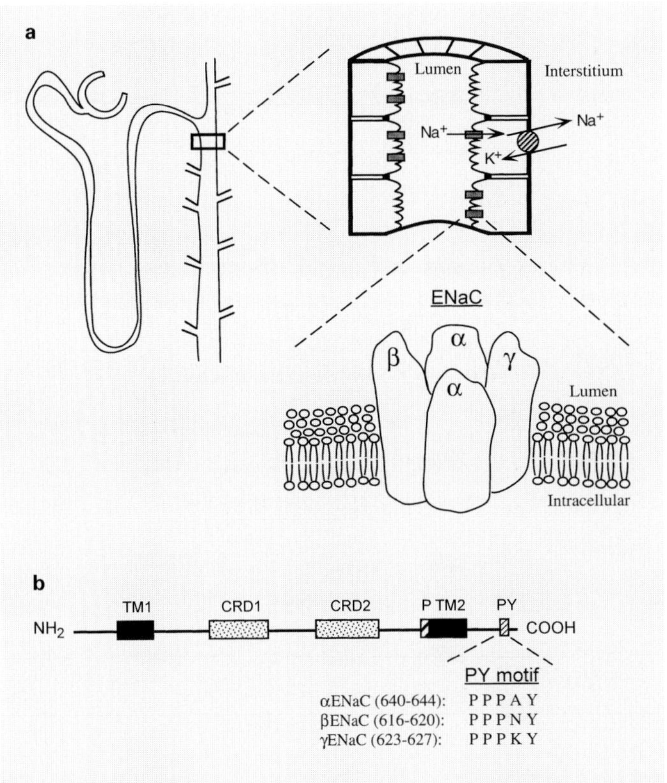

Fig. 1. a Schematic illustration showing the localization of epithelial Na^+ channels (ENaCs) in the collecting duct of the kidney. ENaC is expressed in the principal cells of the collecting duct, and its presence is polarized at the apical membrane in the cell. Na^+ enters the cell passively through ENaC following its electrochemical concentration gradient which is set across the apical membrane. Na^+ leaves the cell by way of Na^+/K^+ ATPase localized in the basolateral membrane. The subunit stoichiometry of the ENaC complex has been suggested either to be 4 subunits (2α, 1β, and 1γ) [14, 15] or 9 subunits (3α, 3β, and 3γ) [16, 17]. Here, a model of the heterotetrameric architecture of ENaC is shown. **b** The three ENaC subunits (α, β, and γ) have a common structure, containing two transmembrane domains (TMs) with a large extracellular loop (about 65% of total amino acids) and two short cytoplasmic amino and carboxy termini. In the extracellular region, there are two conserved cysteine-rich domains (CRDs), which seem to play an important role in channel assembly and targeting [47]. In each ENaC subunit, the pore-forming region (P) is localized just before TM2 [49], and the PY motif (PPPXY; X denotes any amino acid) is localized in the carboxy terminus. Numbers in parentheses correspond to the amino acid positions of the PY motif in each human ENaC subunit. The PY motif is shown to regulate channel density in the plasma membrane by interacting with the WW domains of Nedd4 protein [41, 50]. The PY motif in αENaC has also been suggested to interact with the SH3 domain of α-spectrin for proper targeting of the ENaC complex to the apical membranes [51].

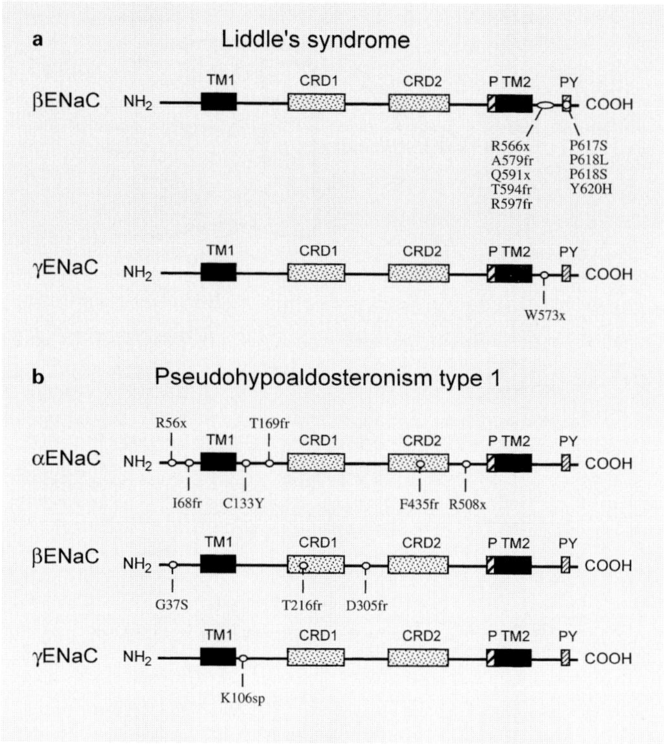

Fig. 2. ENaC mutations identified in Liddle's syndrome (**a**) and autosomal recessive pseudohypoaldosteronism type 1 (PHA-1) (**b**). Liddle's syndrome, an autosomal dominant form of low-renin hypertension, is caused by gain-of-function mutations in the carboxy terminus of the β or γ ENaC subunits that constitutively activate ENaC function. In contrast, loss-of-function mutations in the α, β, and γ ENaC subunits have been found in autosomal recessive PHA-1 [43–45]. The mutated amino acids in Liddle's syndrome or PHA-1 have been numbered according to human ENaC sequences. 'x' denotes a truncated protein caused by a new stop codon that is generated by point mutation. 'fr' denotes a frame shift mutation. 'sp' denotes splice-site mutation. In Liddle's syndrome, R566x is described by Shimkets et al. and Melander et al. [31]; A579fr by Jeunemaitre et al. [32]; Q589x by Shimkets et al. [30]; T594fr by Shimkets et al. [30]; R597fr by Shimkets et al. [30], Jackson et al. [33], and Inoue et al. [34]; P617S by Inoue et al. [35]; P618L by Hansson et al. [36] and Uehara et al. [39]; P618S by Uehara et al. [39]; Y620H by Tamura et al. [38]; W573x by Hansson et al. [37], respectively. In PHA-1, R56x is described by Chang et al. [43] and Kerem et al. [46]; I68fr by Chang et al. [43]; C133Y by Grunder et al. [44]; T169fr by Kerem et al. [46]; F435fr by Kerem et al. [46]; R508x by Chang et al. [43] and Kerem et al. [46]; G37S by Chang et al. [43]; T216fr and D305fr by Chang et al.[43] and Kerem et al. [46]; K106sp by Strautnieks et al. [45], respectively.

Two other ENaC subunits (β and γ subunits) can potentiate the amiloride-sensitive current by more than 20-fold when all three subunits are coexpressed together in *Xenopus* oocytes [3, 18]. Interestingly, neither the ENaC β and γ subunit alone could generate significant amiloride-sensitive currents, which is consistent with an unique role of the α subunit in forming and delivering of the assembled channel complex to the membrane. Moreover, coexpression of α and β subunits (or α and γ subunits) generated a small (3- to -5-fold) or no potentiation in current [3, 18], suggesting that maximal ENaC activity can only be obtained when all three subunits are coexpressed together. Recently, Bonny et al. [23] have described a truncated αENaC which lacks the critical pore-forming (P) region (fig. 1), but is still capable with extended in vitro incubation of delivering a functional amiloride-sensitive channel, presumably composed of truncated αENaC and wild-type β and γ subunits, to the surface membrane in oocytes. The findings reported by Bonny et al. [23] suggest that the truncated αENaC, which retains the N-terminus and the extracellular cysteine-rich domain, may contain a sorting signal that promotes the assembly and intracellular trafficking of the assembled ENaC complexes to the cell membrane.

Although ENaCs of different species are highly homologous to each other, there are some minor functional differences between different species. For example, Chraibi and Horisberger [26] have reported that human and *Xenopus* ENaCs, but not rat ENaCs, can be activated by gilbenclamide, an inhibitor of sulfonylurea receptor and cystic fibrosis conductance regulator, in the oocyte expression system. The amiloride-sensitivity current obtained with α + β subunits of each species is also quite different. Rat α + β ENaC showed 5–30% of the amiloride-sensitive current compared to wild type α + β + γ subunits [27]. In contrast, extremely low amiloride-sensitive currents were observed with human α + β ENaC subunits after expression in the oocyte system [18]. Moreover, our laboratory has recently demonstrated that human ENaC can be activated by a membrane-permeable cAMP analog, 8-CPT-cAMP, while the effect on rat ENaC activity is transient and can be easily missed, especially in the flow-through perfusion system (Tamba et al., unpublished data). These findings emphasize the necessity of using human ENaCs when investigating the structure-function relationship of human disease-related ENaC mutations in the heterologous expression system.

Liddle's Syndrome

Liddle's syndrome represents a rare form of low-renin, volume-expanded, salt-sensitive hypertension. It was first described by Liddle et al. [28] in 1963 in a large kindred from Alabama. Liddle's original publication is a classical description of clinical investigation. They did not find any beneficial effect of spironolactone, a mineralocorticoid antagonist, but triamterene (ENaC blocker) did correct the hypertension, renal potassium wasting, and hypokalemia as long as the subjects restricted their dietary salt intake; salt restriction is necessary because low sodium in the filtered fluid in the kidney can enhance the blocking effect of triamterene on ENaCs. Liddle et al. [28] correctly interpreted these findings as indicating an intrinsic renal defect in the regulation of salt absorption rather than the effects of some unidentified mineralocorticoid other than aldosterone on the distal tubules. In 1994, the autosomal dominant trait of this syndrome was revealed by Botero-Velez et al. [29], and subsequently ENaC mutations associated with the original Liddle's kindred as well as with four other families have been demonstrated by Shimkets et al. [30] using genetic linkage analysis. Several other ENaC mutations associated with Liddle's syndrome have since been identified [31–39]. Interestingly, all of the identified Liddle's syndrome mutations reside in the carboxy terminus of the β or γ ENaC subunits, respectively, either deleting or modifying the functionally important PY motif (fig. 2). Deletion or modification of the PY motif in Liddle's syndrome results in an increased channel number (as well as an increased open probability of the single ion channel) in expression systems [40–42]. Constitutively activated ENaCs in Liddle's syndrome would provide a pathophysiological underlying cause for this rare form of low-renin hypertension. Unregulated, continuing Na^+ reabsorption across the collecting tubules of the kidney would result in volume expansion, suppression of plasma renin activity and aldosterone secretion, and consequent hypertension (fig. 3).

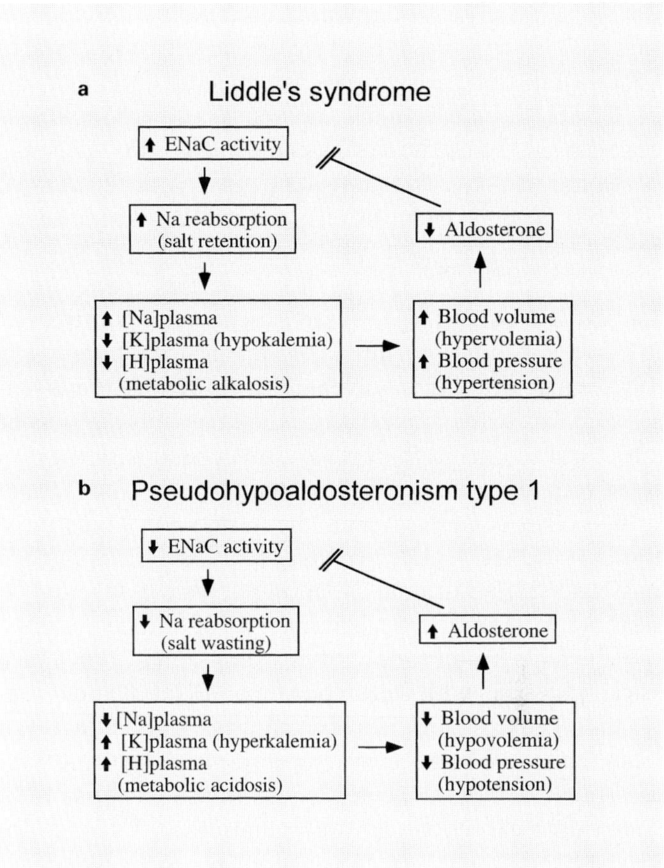

Fig. 3. Possible pathways that are involved in Liddle's syndrome (**a**) and autosomal recessive PHA-1 (**b**). In both cases, the 'dysregulation' of ENaC activity (enhanced channel activity in Liddle's syndrome or reduced channel activity in PHA-1) underlies the pathophysiology.

Pseudohypoaldosteronism Type 1

Genetic linkage analysis has also demonstrated an association of ENaC gene mutations with autosomal recessive pseudohypoaldosteronism type 1 (PHA-1), an inherited human disorder characterized by salt-wasting, hyperkalemic acidosis, and unresponsiveness to mineralocorticoids [43–45]. Premature truncations of αENaC occurring prior to the first transmembrane domain (I68fr) or before the second transmembrane domain (R508x) have been found (fig. 2). Expression of these mutants in *Xenopus* oocytes generated no appreciable level of amiloride-sensitive current [43], which is consistent with the pathophysiology ('loss-of-channel function') of PHA-1. In γENaC, splice-site mutations occurring immediately after the first transmembrane domain (K106sp) have been found, which result in either a premature truncation or substitution of three conserved amino acids of 106–108KYS by an asparagine [45]. Recently, Kerem et al. [46] reported several PHA-1 mutations causing premature truncation of α or β ENaC by point mutation or frame shift (fig. 2) The functional consequences of these mutations have not been reported, but it seems likely that they will generate reduced ENaC activity.

Two single amino acid point mutations have been found in PHA-1 (fig. 2). One mutation occurs at the extracellular region of αENaC (C133Y), which replaces cysteine to tyrosine [44]. In each ENaC subunit, there are two cysteine-rich domains (CRDs) at the extracellular region (fig. 2). Recently, Firsov et al. [47] performed mutational analysis of CRDs in rat ENaC, and found that certain cysteines in CRD1 and CRD2, respectively, are involved in forming a disulfide bridge within each domain. It appears that C133 may form a disulfide bridge with a cysteine in CRD1 (in this regard, we may also consider C133 as a part of CRD1 although there are about 100 amino acids between C133 and the first cysteine in CRD1 of αENaC). Mutation of the C133-equivalent cysteine in rat αENaC (C158S) produced significantly reduced amiloride-sensitive currents in oocytes [47]. Interestingly, this loss-of-function mutation in C158S showed further reduced currents after incubating oocytes at a higher temperature, i.e., 30 °C (the standard incubation temperature for oocytes is 19 °C), suggesting an involvement of the temperature-sensitive ENaC delivery mechanism to the membrane and/or structural stabilization of ENaCs associated with the cysteine mutation. The critical role of CRD in ENaC function is also consistent with the expression study of a human αENaC splice variant, hαENaC+22 [24]. In hαENaC+22, additional 22 amino acids are inserted in frame in CRD2, and negligible levels of amiloride-sensitive currents are produced after expression of this splice variant together with wild-type human ENaC β and γ subunits in oocytes. Another point mutation in PHA-1 has been found in βENaC, replacing a conserved glycine with a serine (G37S) in the cytoplasmic amino terminus [43]. Functional expression of this mutation in *Xenopus* oocytes showed an ENaC activity which was reduced to about 40% of the activity expressing wild-type channels. Grunder et al. [48] have provided evidence that this amino acid may be critically involved in channel gating, i.e., regulation of channel open probability. Reduced ENaC activity in PHA-1 would result in reduced Na^+ reabsorption in the kidney, salt wasting, volume reduction, high plasma potassium level, elevated plasma renin activity and aldosterone level, and consequent hypotension. PHA-1 shows mineralocorticoid resistance at the target organ level due to the 'dysregulation' of ENaCs (fig. 3).

Conclusion

ENaCs play a major physiological role in the human body to maintain Na^+ homeostasis, blood volume, and blood pressure. Genetic linkage analysis has been used to identify specific ENaC mutations occurring in two familial monogenic diseases, Liddle's syndrome and PHA-1. Functional studies of these mutations in the heterologous expression systems have revealed the underlying pathogenesis of these diseases, and have also helped us to better understand the structure-function relationship of these important gene products that are critically involved in renal sodium handling. Involvement of ENaC genes in other forms of human disease, such as other forms of low-renin hypertension or salt-sensitive hypertension, diabetic nephropathy, and even multifactorial polygenic human essential hypertension, will need to be examined in the future.

Acknowledgments

This work was supported in part by grants from the National Institute of Health: NS-34877 (Y.O.), and DK-19407 and DK-53161 (D.G.W.).

References

1 Garty H, Palmer LG: Epithelial sodium channels: Function, structure, and regulation. Physiol Rev 1997;77:359–396.
2 Benos DJ, Cunnigham S, Baker RR, Beason KB, Oh Y, Smith PR: Molecular characteristics of amiloride sensitive sodium chanels. Rev Physiol Biochem Pharmacol 1992;120:31–113.
3 Canessa CM, Schild L, Buell G, Thorens B, Gautschi I, Horisberger J-D, Rossier BC: Amiloride-sensitive epithelial Na^+ channel is made of three homologous subunits. Nature 1994;367:463–467.
4 Hong K, Driscoll M: A transmembrane domain of the putative channel subunit MEC-4 influences mechanotransduction and neurodegeneration *in C. elegans*. Nature 1994;367:470–473.
5 Chalfie M, Driscoll M, Huang M: Degenerin similarities. Nature 1993;361:504.
6 Waldmann R, Champigny G, Bassilana F, Voilley N, Lazdunski M: Molecular cloning and functional expression of a novel amiloride-sensitive Na^+ channel. J Biol Chem 1995;270:27411–27414.
7 Corey DP, Garcia-Anoveros J: Mechanosensation and the DEG/ENaC ion channels. Science 1996;273:323–324.
8 Waldmann R, Champigny G, Bassilana F, Heurteaux C, Lazdunski M: A proton-gated cation channel involved in acid-sensing. Nature 1997;386:173–176.
9 Lingueglia E, Champigny G, Lazdunski M, Barbry P: Cloning of the amiloride-sensitive FMRFamide peptide-gated sodium channel. Nature 1995;378:730–733.
10 Snyder PM, McDonald FJ, Stokes JB, Welsh MJ: Membrane topology of the amiloride-sensitive epithelial sodium channel. J Biol Chem 1994;269:24379–24383.
11 Renard S, Lingueglia E, Voilley N, Lazdunski M, Barbry P: Biochemical analysis of the membrane topology of the amiloride-sensitive Na^+ channel. J Biol Chem 1994;269:12981–12986.
12 Canessa CM, Merillat A-M, Rossier BC: Membrane topology of the epithelial sodium channel in intact cells. Am J Physiol 1994;267:C1682–C1690.
13 Garcia-Anoveros J, Ma C, Chalfie M: Regulation of *Caenorhabditis elegans* degenerin proteins by a putative extracellular domain. Current Biol 1995;5:441–448.

14 Kosari F, Sheng S, Li J, Mak D-O, Foskett JK, Kleyman TR: Subunit stoichiometry of the epithelial sodium channel. J Biol Chem 1998;273: 13469–13474.
15 Firsov D, Gautschi I, Merillat AM, Rossier BC, Schild L: The heterotetrameric architecture of the epithelial sodium channel (ENaC). EMBO J 1998;17:344–352.
16 Eskandari S, Snyder PM, Kreman M, Zampighi GA, Welsh MJ, Wright EM: Number of subunits comprising the epithelial sodium channel. J Biol Chem 1999;274:27281–27286.
17 Snyder PM, Cheng C, Prince LS, Rogers JC, Welsh MJ: Electrophysiological and biochemical evidence that DEG/ENaC cation channels are composed of nine subunits. J Biol Chem 1998;273:681–684.
18 McDonald FJ, Price MP, Snyder PM, Welsh MJ: Cloning and expression of the β- and γ-subunits of the human epithelial sodium channel. Am J Physiol 1995;268:C1157–C1163.
19 Hummler E, Barker P, Gatzy J, Beermann F, Verdumo C, Schmidt A, Boucher R, Rossier BC: Early death due to defective neonatal lung liquid clearance in αENaC-deficient mice. Nature Genet 1996;12:325–328.
20 Matalon S: Mechanisms and regulation of ion transport in adult mammalian alveolar type II pneumocytes. Am J Physiol 1991;261:C727–C738.
21 Oh YS, Saxena S, Warnock DG: αENaC: Leading the charge. J Clin Invest 1999;104:849–850.
22 Masilamani S, Kim G-H, Mitchell C, Wade JB, Knepper MA: Aldosterone-mediated regulation of ENaC α, β, and γ subunit proteins in rat kidney. J Clin Invest 1999;104:R19–R23.
23 Bonny O, Chraibi A, Loffing J, Jaeger NF, Grunder S, Horisberger J-D, Rossier BC: Functional expression of a pseudohypoaldosteronism type I mutated epithelial Na^+ channel deleting the pore-forming region of its α subunit. J Clin Invest 1999;104:967–974.
24 Tucker JK, Tamba K, Lee Y-J, Shen L-L, Warnock DG, Oh Y: Cloning and functional studies of splice variants of the α-subunit of the amiloride-sensitive Na^+ channel. Am J Physiol 1998;274:C1081–C1089.
25 Thomas C, Auerbach S, Stokes JB, Volk KA: 5′-Heterogeneity in epithelial sodium channel α-subunit mRNA leads to distinct NH_2-terminal variant proteins. Am J Physiol 1998;274: C1312–C1323.
26 Chraibi A, Horisberger J-D: Stimulation of epithelial sodium channel activity by the sulfonylurea gilbenclamide. J Pharmacol Exp Ther 1999;290:341–347.
27 McNicholas CM, Canessa CM: Diversity of channels generated by different combinations of epithelial sodium channel subunits. J Gen Physiol 1997;109:681–692.
28 Liddle GW, Bledsoe T, Coppage WS: A familial renal disorder simulating primary aldosteronism but with negligible aldosterone secretion. Trans Am Assoc Phys 1963;76:199–213.
29 Botero-Velez M, Curtis JJ, Warnock DG: Brief report: Liddle's syndrome revisited. N Engl J Med 1994;330:178–181.
30 Shimkets RA, Warnock DG, Bositis CM, Nelson-Williams C, Hansson JH, Schambelan M, Gill JRJ, Ulick S, Milora RV, Findling JW, Canessa CM, Rossier BC, Lifton RP: Liddle's syndrome: heritable human hypertension caused by mutations in the β subunit of the epithelial sodium channel. Cell 1994;79:407–414.
31 Melander O, Orho M, Fagerudd J, Bengtsson K, Groop P-H, Mattiasson I, Groop L, Hulthen UL: Mutations and variants of the epithelial sodium channel gene in Liddle's syndrome and primary hypertension. Hypertension 1998;31: 1118–1124.
32 Jeunemaitre X, Bassilana F, Persu A, Dumont C, Champigny G, Lazdunski M, Corvol P, Barbry P: Genotype-phenotype analysis of a newly discovered family with Liddle's syndrome. J Hypertens 1997;15:1091–1100.
33 Jackson SN, Williams B, Houtman P, Trembath RC: The diagnosis of Liddle syndrome by identification of a mutation in the β subunit of the epithelial sodium channel. J Med Genet 1998;35:510–512.
34 Inoue T, Okauchi Y, Matsuzaki Y, Kuwajima K, Kondo H, Horiuchi N, Nakao K, Iwata M, Yokogoshi Y, Shintani Y, Bando H, Saito S: Identification of a single cytosine base insertion mutation at Arg 597 of the β subunit of the human epithelial sodium channel in a family with Liddle's disease. Eur J Endo 1998;138: 691–697.
35 Inoue J, Iwaoka T, Tokunaga H, Takamune K, Naomi S, Araki M, Takahama K, Yamaguchi K, Tomita K: A family with Liddle's syndrome caused by a new missense mutation in the β subunit of the epithelial sodium channel. J Clin Endo Metabol 1998;83:2210–2213.
36 Hansson JH, Schild L, Lu Y, Wilson TA, Gautschi I, Shimkets R, Nelson-Williams C, Rossier B, Lifton R: A de novo missense mutation of the β subunit of the epithelial sodium channel causes hypertension and Liddle syndrome, identifying a proline-rich segment critical for regulation of channel activity. Proc Natl Acad Sci USA 1995;92:11495–11499.
37 Hansson JH, Nelson-Williams C, Suzuki H, Schild L, Shimkets R, Lu Y, Canessa C, Iwasaki T, Rossier B, Lifton RP: Hypertension caused by a truncated epithelial sodium channel γ subunit: Genetic heterogeneity of Liddle syndrome. Nature Genet 1995;11:76–82.
38 Tamura H, Shild L, Enomoto N, Matsui N, Marumo F, Rossier B, Sasaki S: Liddle disease caused by a missense mutation of β subunit of the epithelial sodium channel gene. J Clin Invest 1996;97:1780–1784.
39 Uehara Y, Sasaquri M, Kinoshita A, Tsuji E, Kiyose H, Taniguchi H, Noda K, Ideishi M, Inoue J, Tomita K, Arakawa K: Genetic analysis of the epithelial sodium channel in Liddle's syndrome. J Hypertens 1998;16:1131–1135.
40 Firsov D, Schild L, Gautschi I, Merillat A-M, Schneeberger E, Rossier BC: Cell surface expression of the epithelial Na channel and a mutant causing Liddle syndrome: a quantitative approach. Proc Natl Acad Sci USA 1996; 93:15370–15375.
41 Staub O, Dho S, Henry PC, Correa J, Ishikawa T, McGlade J, Rotin D: WW domains of Nedd4 bind to the proline-rich PY motifs in the epithelial Na^+ channel deleted in Liddle's syndrome. EMBO J 1996;15:2371–2380.
42 Snyder PM, Price MP, McDonald FJ, Adams CM, Volk KA, Zeiher BG, Stokes JB, Welsh MJ: Mechanism by which Liddle's syndrome mutations increase activity of a human epithelial Na^+ channel. Cell 1995;83:969–978.
43 Chang SS, Grunder S, Hanukoglu A, Rosler A, Mathew PM, Hanukoglu I, Schild L, Lu Y, Shimkets RA, Nelson-Williams C, Rossier BC, Lifton RP: Mutations in subunits of the epithelial sodium channel cause salt wasting with hyperkalaemic acidosis, pseudohypoaldosteronism type 1. Nature Genet 1996;12:248–253.
44 Grunder S, Chang SS, Lifton R, Rossier BC: PHA-1: A novel thermosensitive mutation in the ectodomain of αENaC. J Am Soc Nephrol 1998;9:35A.
45 Strautnieks SS, Thompson RJ, Gardiner RM, Chung E: A novel splice-site mutation in the γ subunit of the epithelial sodium channel gene in three pseudohypoaldosteronism type 1 families. Nature Genet 1996;13:248–250.
46 Kerem E, Bistritzer T, Hanukoglu A, Hofmann T, Zhou Z, Bennett W, MacLaughlin E, Barker P, Nash M, Quittell L, Boucher R, Knowles MR: Pulmonary epithelial sodium-channel dysfunction and excess airway liquid in pseudohypoaldosteronism. N Engl J Med 1999;341: 156–162.
47 Firsov D, Robert-Nicoud M, Gruender S, Schild L, Rossier BC: Mutational analysis of cystein-rich domains of the epithelial sodium channel (ENaC). J Biol Chem 1999;274:2743–2749.
48 Grunder S, Firsov D, Chang SS, Jaeger NF, Gautschi I, Schild L, Lifton RP, Rossier BC: A mutation causing pseudohypoaldosteronism type 1 identifies a conserved glycine that is involved in the gating of the epithelial sodium channel. EMBO J 1997;16:899–907.
49 Schild L, Schneeberger E, Gautschi I, Firsov D: Identification of amino acid residues in the α, β, and γ subunits of the epithelial sodium channel (ENaC) involved in amiloride block and ion permeation. J Gen Physiol 1997;109:15–26.
50 Goulet CC, Volk KA, Adams CM, Prince LS, Stokes JB, Snyder PM: Inhibition of the epithelial Na^+ channel by interaction of Nedd4 with a PY motif deleted in Liddle's syndrome. J Biol Chem 1998;273:30012–30017.
51 Rotin D, Bar-Sagi D, O'Brodovich H, Merilainen J, Lehto VP, Canessa CM, Rossier BC, Downey GP: An SH3 binding region in the epithelial Na^+ channel (αrENaC) mediates its localization at the apical membrane. EMBO J 1994;13:4440–4450.

Experimental Nephrology

Exp Nephrol 2000;8:326–331

Defective Processing and Trafficking of Water Channels in Nephrogenic Diabetes insipidus

E.J. Kamsteeg P.M.T. Deen C.H. van Os

Department of Cell Physiology, University of Nijmegen, The Netherlands

Key Words

Water channels · Nephrogenic diabetes insipidus · Dominant · Recessive · Oligomerization · Trafficking

Abstract

Nephrogenic diabetes insipidus (NDI) is a disease characterized by the inability of the kidney to concentrate urine upon stimulation with vasopressin. Mutations in the gene for aquaporin-2 (AQP2) are the cause of the autosomal recessive and autosomal dominant forms of NDI. Mutant AQP2 proteins, found in autosomal recessive NDI, were shown to be misfolded and retarded in the endoplasmic reticulum. One mutant protein leading to autosomal dominant NDI, E258K, has been analyzed in detail. It was shown that this mutant was not retarded in the endoplasmic reticulum but mainly retained in the Golgi network. Furthermore, this particular mutant was able to form heterotetramers with wild-type AQP2, in contrast to mutants found in autosomal recessive NDI. The subsequent misrouting of complexes containing wild-type and mutant AQP2 proteins explains dominant NDI.

Introduction

Although water can pass lipid bilayer membranes freely by diffusion, the need of bulk water flow through selective channels, called aquaporines (AQPs), is well established to account for the fast and selective transport of water over membranes in specialized tissues. Therefore, it is not surprising that in kidneys, where in a human adult ~180 litres of water are reabsorbed daily, these AQPs play a major role in water reabsorption. AQP1 is essential for reabsorption of water in the proximal tubule and descending limb of Henle's loop [1, 2] and is constitutively present at the basolateral and apical membranes of these cells. In the principal cells of the collecting duct, the antidiuretic hormone arginine vasopressin (AVP) regulates water reabsorption by shuttling of AQP2 between intracellular vesicles and the apical plasma membrane [3]. After passing the apical membrane of the collecting duct via AQP2, water flows through the basolateral membranes via AQP3 and AQP4 [4, 5].

The shuttling of AQP2 is initiated by binding of AVP to the vasopressin type-2 receptor (V2R) at the basolateral side of the collecting duct principal cells. This binding leads, via a cascade of reactions, to an increase in intracellular cAMP levels, resulting in phosphorylation of, among other proteins, AQP2 by protein kinase A. The phosphorylation of AQP2 is essential for translocation and fusion

KARGER

Fax + 41 61 306 12 34
E-Mail karger@karger.ch
www.karger.com

1018–7782/00/0086–0326$17.50/0

Accessible online at:
www.karger.com/journals/exn

Carel H. van Os
Department of Cell Physiology, University of Nijmegen
PO Box 9101, NL–6500 HB Nijmegen (The Netherlands)
Tel. +31 24 3613654, Fax +31 24 3540525
E-Mail C.vanOs@celfys.kun.nl

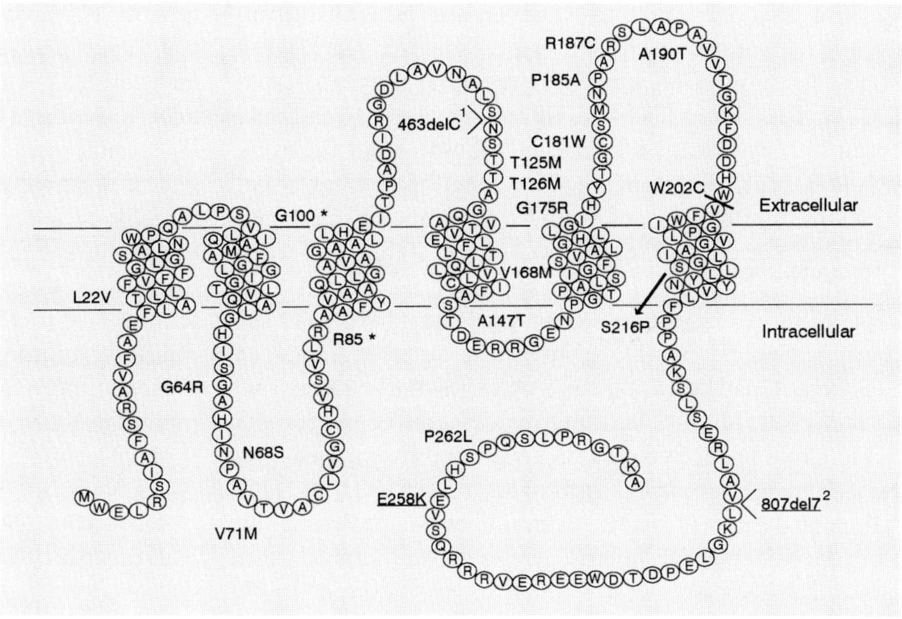

Fig. 1. Proposed topology of the AQP2 water channel. The encoded mutations (amino acid changes, deletions (del) and introduced stopcodons (*)) found in patients with autosomal NDI and their position in the topology are indicated. Mutations identified in dominant NDI are underlined. [1] Deletion of a C at nucleotide position 463 results in a frame shift and premature termination of the protein eight amino acids downstream from the mutation. [2] Deletion of seven nucleotides, from position 807, results in a frame shift and extension of the protein by 60 amino acids.

of intracellular vesicles containing AQP2 with the apical plasma membrane, rendering the cell water permeable. Upon removal of AVP, AQP2 is internalized by endocytosis, which restores the water impermeable state of the cell [3, 6–8].

Nephrogenic diabetes insipidus (NDI) is a disease characterized by the inability of the kidney to concentrate urine upon stimulation by AVP and can be caused by mutations either in the *V2R* gene or in the *AQP2* gene. Mutations in the *V2R* gene cause the X-linked form of NDI [9, 10] which accounts for 90% of the patients with congenital NDI, whereas mutations in the *AQP2* gene are the cause of the autosomal recessive and autosomal dominant forms of inheritance of NDI [11–21]. In this review the defective processing and trafficking of mutant AQP2 proteins, leading to NDI, will be discussed.

Molecular Structure of AQP2

Hydrophobicity plots have predicted that AQPs are integral membrane proteins with six transmembrane domains and five connecting loops (A–E), having the amino and carboxyl termini located intracellularly (fig. 1). This prediction has been confirmed for AQP1 by cryo-electron microscopy [22, 23]. The highly conserved Asp-Pro-Ala (NPA) sequence, which is characteristic for the MIP family to which AQPs belong, are found in loops B and E, which are postulated to fold back into the membrane to form the actual water pore [24]. AQP1 has been shown to form homotetramers [22, 24–26], while the individual monomers are functional water pores [27, 28]. In biological membranes, also AQP2 is found to be present as homotetramers[29]. Due to the similarities of these two proteins, it is very likely that also AQP2 monomers within the tetramers are individual water pores. In other words, the insertion of one AQP2 tetramer into the plasma membrane results in four individual, though connected, water pores.

Defective Processing and Trafficking of AQP2 in Autosomal Recessive NDI

In patients with autosomal recessive NDI, 19 mutations in the AQP2 gene have been reported, leading to mutant AQP2 proteins (fig. 1) [11–19]. Three of these mutations lead to an early stop codon, resulting in truncated proteins (R85*, G100* and 369delC; fig. 1) [12, 16, 18]. One mutation, a G–T transition at nucleotide 700, might result in the substitution of a cysteine for a tryptophan at amino acid 202 (W202C) or an altered splicing (fig. 1) [14]. Ten remaining mutants have been functionally tested in *Xenopus laevis* oocytes, which revealed that seven mutants (G64R, N68S, T125M, G175R, C181W, R187C and S216P; fig. 1) failed to increase water permeability [11, 15, 17, 19, 30]. Realizing that the N68S, C181W and R187C mutants are localized within or close to the conserved NPA boxes, which are thought to form the actual water pore, it is not surprising that they fail to confer to water permeability. Three other mutants tested (L22V, T126M and A147T; fig. 1) displayed a relatively small but significant increase in the water permeability [15, 17], indicating that they are functional water channels. Immunoblotting of all these mutants revealed that the A147T and S216P mutants were less stable than wild-type AQP2, and that, more importantly, all mutants (except for the T125M) showed not only the native ~29 kD band, like wild-type AQP2, but also an additional band of ~32 kD. This additional band was shown to be sensitive to endoglycosidase-H treatment and, therefore, represents a high mannose glycosylated, presumably endoplasmic reticulum (ER)-retarded form of AQP2 [17, 30, 31]. This ER retardation of AQP2 mutants was confirmed by immunofluorescence microscopy, because these mutants showed a dispersed expression throughout the cell [17, 30]. The absence of glycosylated forms of T125M is explained by interference of this mutation with the consensus site for glycosylation in human AQP2 (Asp-Ser-Thr at position 123–125) [19].

Goji et al. [19] suggested that the T125M and G175R mutants are properly routed but inactive water channels, which would explain recessive NDI. However, the proper routing of these mutants is questionable because they conclude that no ER retardation on immunoblot or with fluorescence microscopy was detected for the three mutants tested (T125M, T126M and G175R), whereas other groups [17, 31] clearly observed ER retardation for the T126M mutant. Besides, in another paper the same group did find ER-retardation of the T126M mutant upon expression in LL-CPK$_1$ cells [32]. Additionally, the authors failed to conclude from their own data [19] that there is a ~32-kD protein band visible for the G175R mutant and that all three mutants are, related to the expression in total membranes, expressed at the plasma membrane to a lower extent than wild-type AQP2. These missed observations are strong indications that these mutants are impaired intracellularly. Thus, although attractive as an alternative explanation for NDI, the existence of inactive, properly routed water channels awaits confirmation by other investigators.

The ER retardation of the L22V, T126M, R187C and S216P mutants, found in recessive NDI, was confirmed by expression in mammalian cells [31, 32], whereas both reports are contradictory for the A147T mutant. In CHO cells, it was shown by [^{35}S] pulse chase that L22V and T126M were as stable as wild-type AQP2, whereas A147T, C181W and R187C were less stable than wild-type AQP2. It was concluded, however, that the decreased stability is unlikely to be the defect causing NDI. Concomitantly, ER retardation of mutant AQP2 proteins is now broadly accepted as the major cause underlying autosomal recessive NDI.

For a growing number of proteins, it has been shown that ER-retarded transport is caused by the inability of the protein to fold into its native structure and is usually followed by degradation in the ubiquitin-proteasome system [33–35]. Therefore, it is striking that all AQP2 mutants found in recessive NDI (except P262L) are present inbetween the first and last transmembrane domain (fig. 1).

Defective Trafficking of AQP2 in Autosomal Dominant NDI

So far, two mutations in the AQP2 gene, causing autosomal dominant NDI (E258K [20] and 807del7 [13]; fig. 1) have been described. Upon expression in oocytes it was shown that the E258K mutant was not retarded in the ER, but was mainly localized to the Golgi complex. The reason why this mutant is not transported to the plasma membrane is not known, but it is suggested that a Golgi retention signal has been introduced by the charge change, inherent to this mutation (E^- to K^+). Osmotic swelling measurements revealed that E258K conferred water permeability to oocytes, although to a lower extent than wild-type AQP2. In AQP2, phosphorylation of the serine at position 256 (S256) by protein kinase A plays an important role in the shuttling of AQP2-containing vesicles. The E258K mutation and S256 are only two amino acids apart and, therefore, it was hypothesized that the E258K mutation would interfere with this phosphorylation. However, an in vitro phosphorylation study showed that the E258K mutant was phosphorylated to the same extent as wild-type AQP2. The water permeability, observed in oocytes expressing the E258K mutant, could be an effect of high expression levels, leading to the saturation of the oocyte's organelles for this mutant, consequently leading to an aberrant localization at the plasma membrane. Indeed, at lower expression levels the oocytes expressing the E258K mutant did not show an increase in water permeability, whereas oocytes expressing the same amount of wild-type AQP2 exhibited water permeability that was five times higher than that of control oocytes. At this low expression level the dominant negative character of E258K could be confirmed [20]. The observed dominant 'loss of function' can only be explained if E258K is

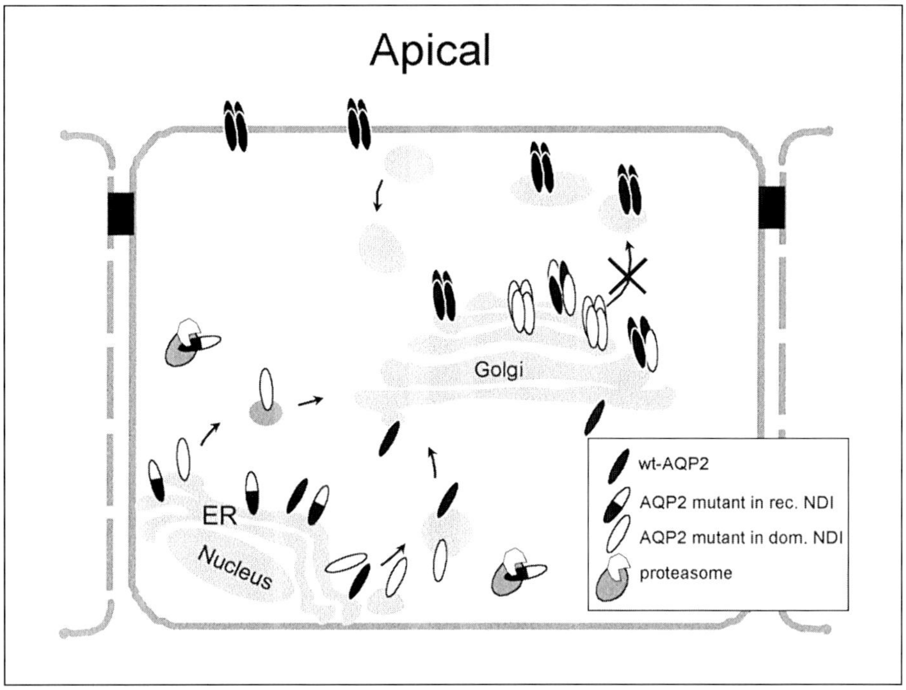

Fig. 2. Schematic representation of the postulated molecular mechanisms underlying autosomal recessive and dominant NDI. Wild-type AQP2 (wt-AQP2) is synthesized in the ER, transported to the Golgi complex and endosomes, and is available for shuttling to the apical membrane. AQP2 mutants in recessive NDI (rec. NDI) are not properly folded, retarded in the ER and possibly degraded by proteasomes. AQP2 mutants in dominant NDI (dom. NDI) are not retarded in the ER, able to oligomerize with wild-type AQP2 and impair the routing of these complexes to the apical membrane (e.g. by impairment of Golgi to apical membrane transport as found for E258K/wild-type AQP2 complexes).

able to form tetramers with wild-type AQP2, thereby inhibiting the function and routing of this complex. As predicted, it has been shown that E258K, in contrast to a mutant found in recessive NDI (R187C), was able to form tetramers and, more importantly, to form heterotetramers with wild-type AQP2 [29]. The finding that the ER-retarded mutant R187C, found in recessive NDI, was not able to form tetramers, suggests that in oocytes, AQP2 assembly occurs in a post-ER compartment which has been shown for connexin-43 [36] and influenza hemagglutinin [37]. Alternatively, compartmentalization of recessive mutants in the ER, resulting in the inability to oligomerize with wild-type AQP2 also explains the observed data.

In the other family with autosomal dominant NDI, a deletion of 7 basepairs from position 807 (807del7) was observed, leading to a 180-basepair-longer transcript. The resulting AQP2 mutant has a frame shift following valine at position 236 and an extended carboxy terminus of 60 amino acids. When expressed in oocytes, the 807del7 mutant showed no enhancement of water permeability but the dominant character could not be confirmed [13]. The formation and subsequent misrouting of mutant/wild-type AQP2 heterotetramers is the only explanation for dominant NDI and will also be applicable to the 807del7 mutant and yet unidentified mutants causing dominant NDI.

Correction of Trafficking of AQP2 Mutants

Newly synthesized proteins in the secretory pathway undergo modification and folding in the ER, which possesses a quality-control mechanism that is directed by molecular chaperones [38]. Misfolded proteins can have a prolonged association with the chaperone often leading to targeting of the protein for degradation [39]. The delF508 mutant of the cystic fibrosis transmembrane conductance regulator is a misfolded protein targeted for degradation. In cell culture models, the mistrafficking of this mutant has been shown to be partially correctable by chemical chaperones [40, 41]. The accumulation of AQP2 mutants in the ER, as has been observed in recessive NDI, suggests that mistrafficking of these mutants is also caused by defective folding. In CHO cells, the chemical chaperones glycerol, TMAO and DMSO have been shown to promote translocation of AQP2 mutants from the ER to the endosomal/membrane compartment, and thus partially correct the mistrafficking of AQP2 proteins [31]. Although the attainability of the use of these chemicals in humans is not known, it suggests a pharmacological strategy for in vivo correction of defective AQP2 processing in NDI.

Infusion of dDAVP, a V2R agonist, showed a transient increase in urinary osmolality of the NDI patients carrying the E258K mutation, whereas the urine osmolality of patients with autosomal recessive NDI did not change

[20]. Assuming equal expression levels of wild-type AQP2 and AQP2-E258K, this suggests that in vivo trafficking of E258K to the plasma membrane can occur, but that retention in the Golgi is the steady-state situation. Therefore, continuous administration of dDAVP to these patients might partially relieve the burden of polyuria and polydipsia, the most conspicious symptoms of NDI.

Conclusions

On the basis of the papers discussed, we can conclude that the changed subcellular localization of the respective AQP2 mutants provides the explanation for the mechanisms involved in recessive and dominant NDI (fig. 2). Except, perhaps, for the mutants studied by Goji et al. [19], all the AQP2 mutants studied in recessive NDI have been shown to be ER-retarded. This ER retardation, which has been found for many other proteins in recessive diseases, is a result of the quality control retention mechanism in the ER for misfolded proteins and is usually followed by degradation of these proteins by cytosolic proteasomes [39]. One of the mutants in dominant NDI (E258K) is not recognized as misfolded by the ER quality control and is as stable as wild-type AQP2 but is retained in the Golgi complex [20]. In recessive NDI, the AQP2 mutants are trapped in the ER and are not able to oligomerize with wild-type AQP2. Consequently, in the healthy parents of patients with a recessive form of NDI, whose genome encodes mutant and wild-type-AQP2 proteins, wild-type-AQP2 proteins are transported from the ER, can form homotetramers and fulfill their role in the concentration of urine. The genome of the patient with dominant NDI also encodes a mutant and a wild-type AQP2 protein. However, this dominant mutant exits the ER, after which it can heterotetramerize with wild-type AQP2. Retardation of the dominant mutant in a post-ER compartment, like the Golgi for the E258K mutant, entraps the heterotetramer in this cell organelle. This precludes routing of sufficient wild-type AQP2 proteins to the apical membrane of collecting duct cells to concentrate urine properly, resulting in NDI.

References

1 Schnermann J, Chou CL, Ma T, Traynor T, Knepper MA, Verkman AS: Defective proximal tubular fluid reabsorption in transgenic aquaporin-1 null mice. Proc Natl Acad Sci USA 1998;95:9660–9664.

2 King LS, Agre P: Pathophysiology of the aquaporin water channels. Annu Rev Physiol 1996; 58:619–648.

3 Nielsen S, Chou CL, Marples D, Christensen EI, Kishore BK, Knepper MA: Vasopressin increases water permeability of kidney collecting duct by inducing translocation of aquaporin-CD water channels to plasma membrane. Proc Natl Acad Sci USA 1995;92:1013–1017.

4 Ecelbarger CA, Terris J, Frindt G, Echevarria M, Marples D, Nielsen S, Knepper MA: Aquaporin-3 water channel localization and regulation in rat kidney. Am J Physiol 1995;38: F663–F672.

5 Terris J, Ecelbarger CA, Marples D, Knepper MA, Nielsen S: Distribution of aquaporin-4 water channel expression within rat kidney. Am J Physiol 1995;38:F775–F785.

6 Marples D, Knepper MA, Christensen EI, Nielsen S: Redistribution of aquaporin-2 water channels induced by vasopressin in rat kidney inner medullary collecting duct. Am J Physiol 1995;38:C655–C664.

7 Katsura T, Ausiello DA, Brown D: Direct demonstration of aquaporin-2 water channel recycling in stably transfected LLC-PK1 epithelial cells. Am J Physiol 1996;39:F548–F553.

8 Yamamoto T, Sasaki S, Fushimi K, Ishibashi K, Yaoita E, Kawasaki K, Marumo F, Kihara I: Vasopressin increases AQP-CD water channel in apical membrane of collecting duct cells in Brattleboro rats. Am J Physiol 1995;268: C1546–C1551.

9 Rosenthal W, Seibold A, Antaramian A, Lonergan M, Arthus M-F, Hendy GN, Birnbaumer M, Bichet DG: Molecular identification of the gene responsible for congenital nephrogenic diabetes insipidus. Nature 1992;359:233–235.

10 van den Ouweland AM, Dreesen JC, Verdijk MAJ, Knoers NVAM: Mutations in the vasopressin type 2 receptor gene (AVPR2) associated with nephrogenic diabetes insipidus. Nature Genet 1992;2:99–102.

11 Deen PMT, Verdijk MAJ, Knoers NVAM, Wieringa B, Monnens LAH, van Os CH, van Oost BA: Requirement of human renal water channel aquaporin-2 for vasopressin-dependent concentration of urine. Science 1994;264: 92–95.

12 van Lieburg AF, Verdijk MAJ, Knoers NVAM, van Essen AJ, Proesmans W, Mallmann R, Monnens LAH, van Oost BA, van Os CH, Deen PMT: Patients with autosomal nephrogenic diabetes insipidus homozygous for mutations in the aquaporin 2 water-channel gene. Am J Hum Genet 1994;55:648–652.

13 Bichet DG, Arthus M-F, Lonergan M, Balfe W, Skorechi K, Nivet H, Robertson G, Oksche A, Rosenthal W, Fujiwara M, Morgan K, Sasaki S: Autosomal dominant and autosomal recessive nephrogenic diabetes insipidus: Novel mutations in the AQP2 gene (abstract). J Am Soc Nephrol 1995;6:717.

14 Oksche A, Moller A, Dickson J, Rosendahl W, Rascher W, Bichet DG, Rosenthal W: Two novel mutations in the aquaporin-2 and the vasopressin V2 receptor genes in patients with congenital nephrogenic diabetes insipidus. Hum Genet 1996;98:587–589.

15 Canfield MC, Tamarappoo BK, Moses AM, Verkman AS, Holtzman EJ: Identification and characterization of aquaporin-2 water channel mutations causing nephrogenic diabetes insipidus with partial vasopressin response. Hum Mol Genet 1997;6:1865–1871.

16 Hochberg Z, van Lieburg AF, Even L, Brenner B, Lanir N, van Oost BA, Knoers NVAM: Autosomal recessive nephrogenic diabetes insipidus caused by an aquaporin-2 mutation. J Clin Endocrinol Metab 1997;82:686–689.

17 Mulders SM, Knoers NVAM, van Lieburg AF, Monnens LAH, Leumann E, Wuhl E, Schober E, Rijss JPL, van Os CH, Deen PMT: New mutations in the AQP2 gene in nephrogenic diabetes insipidus resulting in functional but misrouted water channels. J Am Soc Nephrol 1997;8:242–248.

18 Vargas-Poussou R, Forestier L, Dautzenberg MD, Niaudet P, Dechaux M, Antignac C: Mutations in the vasopressin V2 receptor and aquaporin-2 genes in twelve families with congenital nephrogenic diabetes insipidus. Adv Exp Med Biol 1998;449:387–390.
19 Goji K, Kuwahara M, Gu Y, Matsuo M, Marumo F, Sasaki S: Novel mutations in aquaporin-2 gene in female siblings with nephrogenic diabetes insipidus: evidence of disrupted water channel function. J Clin Endocrinol Metab 1998;83:3205–3209.
20 Mulders SM, Bichet DG, Rijss JPL, Kamsteeg EJ, Arthus MF, Lonergan M, Fujiwara M, Morgan K, Leijendekker R, van der Sluijs P, van Os CH, Deen PMT: An aquaporin-2 water channel mutant which causes autosomal dominant nephrogenic diabetes insipidus is retained in the Golgi complex. J Clin Invest 1998;102: 57–66.
21 Kuwahara M: Aquaporin-2, a vasopressin-sensitive water channel, and nephrogenic diabetes insipidus. Intern Med 1998;37:215–217.
22 Walz T, Hirai T, Murata K, Heymann JB, Mitsuoka K, Fujiyoshi Y, Smith BL, Agre P, Engel A: The three-dimensional structure of aquaporin-1. Nature 1997;387:624–627.
23 Walz T, Tittmann P, Fuchs KH, Muller DJ, Smith BL, Agre P, Gross H, Engel A: Surface topographies at subnanometer-resolution reveal asymmetry and sidedness of aquaporin-1. J Mol Biol 1996;264:907–918.
24 Jung JS, Preston GM, Smith BL, Guggino WB, Agre P: Molecular structure of the water channel through aquaporin CHIP. The hourglass model. J Biol Chem 1994;269:14648–14654.
25 Smith BL, Agre P: Erythrocyte Mr 28,000 transmembrane protein exists as a multisubunit oligomer similar to channel proteins. J Biol Chem 1991;266:6407–6415.
26 Verbavatz JM, Brown D, Sabolic I, Valenti G, Ausiello DA, Van Hoek AN, Ma T, Verkman AS: Tetrameric assembly of CHIP28 water channels in liposomes and cell membranes: A freeze-fracture study. J Cell Biol 1993;123: 605–618.
27 Van Hoek AN, Hom ML, Luthjens LH, de Jong MD, Dempster JA, van Os CH: Functional unit of 30 kDa for proximal tubule water channels as revealed by radiation inactivation. J Biol Chem 1991;266:16633–16635.
28 Preston GM, Jung JS, Guggino WB, Agre P: The mercury-sensitive residue at cysteine 189 in the CHIP28 water channel. J Biol Chem 1993;268:17–20.
29 Kamsteeg EJ, Wormhoudt TA, Rijss JPL, van Os CH, Deen PMT: An impaired routing of wild-type aquaporin-2 after tetramerization with an aquaporin-2 mutant explains dominant nephrogenic diabetes insipidus. EMBO J 1999;18:2394–2400.
30 Deen PMT, Croes H, van Aubel RA, Ginsel LA, van Os CH: Water channels encoded by mutant aquaporin-2 genes in nephrogenic diabetes insipidus are impaired in their cellular routing. J Clin Invest 1995;95:2291–2296.
31 Tamarappoo BK, Verkman AS: Defective aquaporin-2 trafficking in nephrogenic diabetes insipidus and correction by chemical chaperones. J Clin Invest 1998;101:2257–2267.
32 Yamauchi K, Fushimi K, Yamashita Y, Shinbo I, Sasaki S, Marumo F: Effects of missense mutations on rat aquaporin-2 in LLC-PK1 porcine kidney cells. Kidney Int 1999;56:164–171.
33 Thomas PJ, Qu BH, Pedersen PL: Defective protein folding as a basis of human disease. Trends Biochem Sci 1995;20:456–459.
34 Hobbs HH, Russell DW, Brown MS, Goldstein JL: The LDL receptor locus in familial hypercholesterolemia: Mutational analysis of a membrane protein. Annu Rev Genet 1990;24:133–170.
35 Hochstrasser M: Protein degradation or regulation: Ub the judge. Cell 1996;84:813–815.
36 Musil LS, Goodenough DA: Multisubunit assembly of an integral plasma membrane channel protein, gap junction connexin43, occurs after exit from the ER. Cell 1993;74:1065–1077.
37 Tatu U, Hammond C, Helenius A: Folding and oligomerization of influenza hemagglutinin in the ER and the intermediate compartment. EMBO J 1995;14:1340–1348.
38 Hammond C, Helenius A: Quality control in the secretory pathway. Curr Opin Cell Biol 1995;7:523–529.
39 Kopito RR: ER quality control: The cytoplasmic connection. Cell 1997;88:427–430.
40 Brown CR, Hong-Brown LQ, Biwersi J, Verkman AS, Welch WJ: Chemical chaperones correct the mutant phenotype of the delta F508 cystic fibrosis transmembrane conductance regulator protein. Cell Stress Chaperones 1996; 1:117–125.
41 Sato S, Ward CL, Krouse ME, Wine JJ, Kopito RR: Glycerol reverses the misfolding phenotype of the most common cystic fibrosis mutation. J Biol Chem 1996;271:635–638.

Experimental Nephrology

Exp Nephrol 2000;8:332–342

Defects in Processing and Trafficking of Cystic Fibrosis Transmembrane Conductance Regulator

Karl Kunzelmann[a] Roland Nitschke[b]

[a]Department of Physiology and Pharmacology, University of Queensland, St. Lucia, Brisbane, Australia, and
[b]Physiologisches Institut, Albert-Ludwigs-Universität Freiburg, Deutschland

Key Words
Cystic fibrosis transmembrane conductance regulator · CFTR · Processing · Traffic · Cystic fibrosis

Abstract
In most epithelial tissues Cl^- transport relies on the cystic fibrosis transmembrane conductance regulator (CFTR) which has dual function as a Cl^- channel and as a regulator of other ion channels. More than 900 different mutations in the CFTR gene are the cause for defective transport of Cl^- and Na^+ and impaired secretion or absorption of electrolytes in cystic fibrosis. However, the CFTR mutation ΔF508 is the most common reason for the frequently inherited disease among the Caucasian population. Maturation and processing of ΔF508-CFTR is defective which leads to expression of only very little but functional CFTR in the cell membrane. Understanding the processing and trafficking of CFTR may give a clue to the question as to how the expression and residual function of ΔF508-CFTR can be enhanced, and may lead to the development of new pharmacological tools for the treatment of cystic fibrosis.

Introduction

The cystic fibrosis transmembrane conductance regulator (CFTR) is essential for epithelial electrolyte transport because the protein forms the apical Cl^- conductance in secretory and also reabsorptive epithelial cells [1, 2]. In addition, CFTR is a regulator of other membrane conductances that participate in transepithelial transport [1, 3, 4]. In this short review only certain aspects of CFTR will be discussed. After a brief introduction into the structure and regulation of expression of CFTR in epithelial cells, we will focus on the process of maturation and defective processing of the most common CFTR mutant, ΔF508. The current knowledge in cellular localization and trafficking of CFTR via exocytosis and endocytosis will also be reviewed together with the recent studies aimed at enhancing the processing and activation of ΔF508-CFTR.

Structure of CFTR

CFTR consists of 1,480 amino acids that form an integral membrane glycoprotein of about 160–180 kD when analyzed by PAGE electrophoresis [5, 6]. CFTR shows

KARGER
Fax + 41 61 306 12 34
E-Mail karger@karger.ch
www.karger.com

1018–7782/00/0086–0332$17.50/0

Accessible online at:
www.karger.com/journals/exn

Karl Kunzelmann
Department of Physiology and Pharmacology
University of Queensland
St. Lucia, Brisbane, Qld. 4072 (Australia)
Tel. +61 7 3375 4104, Fax +61 7 3365 4933, E-Mail kunzelmann@plpk.uq.edu.au

striking similarities to other membrane transport proteins, according to hydropathy analysis. It therefore belongs to the ABC transporter superfamily [5, 7]. These ABC proteins are characterized by a repeat of 6 transmembrane helices (TMD1, 2), two nucleotide-binding folds (NBF1, 2), and a central cytosolic domain. In the case of CFTR, this central cytosolic part of the protein is relatively large and contains numerous phosphorylation sites for protein kinase A (PKA) and protein kinase C (PKC). It was therefore denoted as the regulatory (R) domain [7–10]. The current model of CFTR, however, remains uncertain since all present knowledge is based mainly on hydropathy analysis and on results obtained from experiments using site-directed mutagenesis of CFTR, cysteine scanning and other indirect methods [11–13]. Thus some studies suggested that NBF1 is partially localized within the cell membrane [14–17]. Moreover, it remains uncertain how far NBF1 extends at the N-terminal site [18, 19]. Ultimately, recent efforts in X-ray analysis of partially crystallized CFTR domains will give a better insight into CFTR's true structure [20].

Cystic Fibrosis

Cystic fibrosis (CF) is caused by more than 900 different CFTR mutations described so far (the Cystic Fibrosis Genetic Analysis Consortium). Depending on the mutation, either biosynthesis or function of the CFTR molecule is impaired. Some mutations lead to decreased expression of CFTR (class 5) or completely abolish CFTR expression (class 1). Others inhibit maturation of the CFTR protein (class 2). In addition, some mutations seem to interfere with Cl^- channel regulation (class 3), or affect single channel conductance (class 4). Because CFTR also regulates the function of other ion channels and participates in cellular functions such as endosomal acidification and exocytosis, some mutations may abolish CFTR's ability to control other membrane conductances (class 6) [2, 21–24]. Despite the large number of CFTR mutations, the ΔF508 allele accounts for almost 70% of the CF chromosomes and thus more than 90% of the CF patients are heterozygous for this mutation. Therefore, a better understanding of the mechanism by which this mutation affects the function of CFTR is essential and may ultimately lead to new therapeutic strategies [25].

As pointed out, CFTR affects a variety of cellular processes. However, the question of how CFTR mutations lead to the disease CF has still not been completely answered. It is now quite obvious that both impaired Cl^- secretion as well as enhanced electrolyte reabsorption contribute to the airway disease, although alternative explanations were presented by other groups. These alternative models include pathophysiological concepts such as: (i) failure in protecting the airways from bacterial infections by β-defensins; (ii) CFTR acting as a receptor for *Pseudomonas aeruginosa,* and (iii) oversulfation and overfucosylation of glycoproteins leading to enhanced attachment of pathogenic bacteria like *P. aeruginosa* [1, 2, 26, 27]. Moreover, additional factors like the defect in mucin secretion, dysregulation of water permeability and bicarbonate secretion as well as inflammatory mechanisms probably contribute to the development of the disease. A more detailed discussion of these mechanisms is provided by several recent reviews [1, 2, 28–30]. However, it is important to note that the severity of the disease in a patient is related to the level of expression of functional CFTR and not to these mechanisms.

Expression of CFTR in Epithelial Cells

CF affects epithelial tissues that express CFTR and thus interferes with epithelial transport of electrolytes [1]. CFTR is localized predominantly in the apical membrane of epithelial cells and might also be localized in some basolateral membranes. Thus, defective Cl^- conductance was detected on both sides of polarized epithelia [31–34]. The functional significance of CFTR expression in the kidney and in nonepithelial tissues like heart muscle cells, brain and lymphocytes remains obscure. Because these tissues are not affected in CF, there is no conclusive evidence that CFTR mutations contribute to any kind of pathological process in these organs [35–39]. Expression of CFTR might therefore be functionally irrelevant or the CFTR defect could be compensated by expression of other proteins in these tissues.

Relatively low levels of CFTR are expressed in native epithelial cells [40–42]. Moreover, expression largely depends on cellular differentiation [43–45]. Cultured cells expressing either endogenous or recombinant CFTR generally show much higher levels of CFTR protein. It is possible that processing or function of CFTR are changed in these cells due to overexpression [46–48]. The CFTR promotor region has been characterized [49, 50], and it has been demonstrated that expression of CFTR is controlled by a cAMP response element and that an increase in intracellular cAMP leads to higher levels of CFTR expression [51]. On the other hand, activation of intracellular PKC inhibits expression of CFTR and it has been demonstrated that PKC acts on a post-transcriptional site by controlling translation of CFTR [52–55]. Similar responses have also been reported following changes in intracellular divalent cations [56]. Moreover, induction of the expression of other proteins like the multidrug resistance protein or the epithelial Na^+ channel (EnaC) is correlated to a decrease in CFTR expression [57–59]. However, even more important than transcriptional control is the post-translational control of the maturation of CFTR by the so-called quality control system, located in the endoplasmic reticulum (ER). This is confirmed by the finding that expression of CFTR mRNA and the amounts of cAMP-activated whole cell Cl^- currents often do not correlate well [42].

Processing of Wild-Type and Mutant CFTR

Maturation of CFTR is a rather inefficient process in that most of the newly synthesized CFTR is degraded during processing in the ER. Thus, about 75% of wild-type (wt) CFTR and basically 100% of ΔF508-CFTR is ubiquitinated and degraded by cytosolic proteasomes and other ATP-dependent proteases described below [60–62]. Cells which show a large expression of CFTR also show large amounts of degradation of CFTR, thereby exhausting the proteasomal capacity. This results in accumulation of tightly packed polyubiquitinated CFTR and the formation of so-called aggresomes that are probably associated with centrosomes [63, 64]. Although an initial report showed that CFTR functions as a monomeric protein [65], a subsequent study suggested that a single channel is formed by two CFTR molecules [66]. Thus, limited supplies of assembly factors could be one reason for inefficient maturation of CFTR. In addition, a rapid degradation of wtCFTR (half-life about 7.5 h) and ΔF508-CFTR (about 20–40 min) is an intrinsic property of this protein and is not caused by overexpression [67]. Due to incomplete folding, ΔF508-CFTR is not capable of leaving the ER and thus does not progress to a more mature form and become protease-resistant [6]. The retention of ΔF508-CFTR in the ER prevents complex glycosylation and it is therefore detected as a band of lower molecular weight when analyzed by SDS gel electrophoresis. The defective glycosylation is the result of misfolding and prolonged chaperon interaction and is not itself the primary cause for defective processing of ΔF508-CFTR [68, 69].

Processing of ΔF508-CFTR is a temperature-sensitive process [70, 71]. When ΔF508-CFTR-expressing cells are kept at a low temperature, the function of the quality-control mechanism responsible for the rapid degradation of folding mutants like ΔF508-CFTR is impaired. This results in a higher yield of completely processed mutant CFTR and thus cAMP-activated whole cell Cl^- conductance is enhanced. Because Sf9 cells and also *Xenopus* ooyctes are kept at temperatures lower than 37 °C, whole cell Cl^- currents generated by ΔF508-CFTR can be detected [72, 73]. Functional ΔF508-CFTR can also be detected upon overexpression in mammalian cells due to a limited capacity of the quality-control system [74]. The ΔF508 mutation affects an early step in the process of protein folding [75] and this leads to most of the ΔF508-CFTR being located intracellularly where it is probably attached to the ER and not in the apical membrane like wtCFTR [76, 77]. The ΔF508 mutation is not the only mutation which compromises biosynthetic processing and mutations in the cytosolic loops, the transmembrane domains or the nucleotide-binding domains of CFTR may also affect maturation, eventually in a somewhat different way [78–82]. Interestingly, a mutation corresponding to ΔF508 which has been introduced into the homologous yeast protein Yor1p, also leads to retention of immature protein in the ER [83]. Moreover, additional mutations partially reverting the deleterious effect of ΔF508 have been identified in STE6-CFTR chimeras in yeast [84].

Once mutant ΔF508-CFTR has been allowed to mature and to traffic to the cell membrane, it probably functions as a Cl^- channel just as wtCFTR. An initial report indicating a change in the single-channel kinetics of ΔF508-CFTR remains unconfirmed by subsequent reports [85, 86], and thus, missing whole cell Cl^- conductance in the epithelial tissues of ΔF508/ΔF508 homozygous patients is most likely caused by defective maturation rather than changes in the single channel properties of CFTR. Nevertheless, ΔF508-CFTR that has escaped proteolysis and entered the plasma membrane seems to have a reduced functional half-life at the plasma membrane [87] and is not as stable as wtCFTR.

Mutant CFTR-like ΔF508-CFTR is trapped in the ER and rapidly degraded. Thus ΔF508-CFTR is mislocalized within the cell. Instead of being present in the apical membrane of epithelial cells, low amounts of ΔF508-CFTR are detected in the cytosol located close to the ER. This has been demonstrated initially in the sweat duct of CF patients [88]. CFTR is polyubiquitinated and subsequently degraded by cytosolic proteolytic activity. Several proteolytic systems, including proteasomes, contribute to the degradation of wt- and ΔF508-CFTR [60, 61]. Degradation of CFTR probably takes place in the cytosol and thus misfolded portions of CFTR located in the ER membrane or the ER lumen will be detached from the bilayer and will be retrotranslocated out of the ER and back into the cytosol [89]. It should be mentioned here, that mutant ΔF508-CFTR was demonstrated to be functional when retained in the ER and was claimed to serve as a channel permeable for ATP and phosphoadenosine phosphosulfate [90]. The enzymes and compartments responsible for degradation of CFTR are part of the so-called ER quality-control system of secretory proteins [91]. Binding of CFTR to the ER membrane chaperon calnexin [92] and the cytosolic chaperon Hsp70 [93] has been demonstrated. In contrast, the luminal ER chaperons Bip and GRP94 do not interact with CFTR. Hsp70 is not the only cytosolic chaperon that binds to CFTR [94]. Recently, it was demonstrated that the most abundant cytosolic chaperon, Hsp90, also binds to CFTR during processing [95]. These interesting results have been obtained in transformed cells and the extent to which they are representative of native cells remains to be determined. Thus it is important to note that both wtCFTR and ΔF508-CFTR have been detected in the apical membranes of native respiratory and intestinal epithelial cells [96, 97].

Taken together, there is a large body of evidence that defective maturation of the most common CFTR mutant ΔF508 is the major cause of CF. Therapeutic intervention in order to enhance maturation and expression of ΔF508 should therefore be beneficial to CF patients as described below.

Cellular Localization of CFTR

Mature fully glycosylated wtCFTR is located in the apical membrane of epithelial cells, whereas misfolded CFTR remains intracellular and is rapidly degraded. However, wtCFTR has also been detected in intracellular compartments like endosomal vesicles [98–100]. Moreover, subapical localization of CFTR, e.g. in clathrin-coated vesicles, has been verified using various fluorescence signals, immunogold labelling and functional methods [77, 98, 100–102]. In endosomes, CFTR may serve as a parallel Cl^- conductance counterbalancing the activity and charge movement of the H^+-ATPase, which acidifies the endosomal compartment [27, 103]. Defective CFTR Cl^- conductance in CF would then interfere with acidification in the endosomes and a higher endosomal pH has been found in CF epithelial cells in one study [103]. However, these results have not been confirmed by subsequent studies [104, 105]. The endosomal compartment is a major site for post-translational modification of proteins such as sialylation of glycoproteins, and the activity of the respective enzymes is largely pH-dependent. In agreement with the initial report [103], a reduced activity of sialyltransferase has been detected in CF cells. As a consequence, membrane proteins with reduced sialylation and enhanced sulfation were detected at the cell surface of CF cells [27, 106–109]. However, these observations do not clarify whether the decrease in sialylation is due to a lack of functional CFTR, or whether it is due to the expression of mutant CFTR.

The extent to which these putative changes in intracellular pH interfere with cellular functions and their quantitative contribution to the malfunction in CF is at present not established [110–113]. An oversulfation and overfucosylation of glycoproteins apparently leads to an enhanced attachment of pathogenic bacteria like *P. aeruginosa* [106, 109, 114] and this mechanism is likely to contribute to the high morbidity and mortality seen in CF [109]. However, it should also be mentioned that severe bacterial colonization of CF lungs is also explained by a different mechanism. According to this, CFTR may act as a receptor for *P. aeruginosa* in the airways and for *Salmonella typhi* in the intestine [115, 116]. Bacteria which bind to CFTR are subsequently taken up by the epithelial cells and then degraded in the lysosomal compartments. However, mutant CFTRs, e.g. ΔF508, do not reach the cell membrane and *P. aeruginosa* and *S. typhi* are therefore not bound and cleared from the airways and intestines, respectively.

Exocytosis and Endocytosis of CFTR

The extent to which PKA-dependent activation of CFTR that is present in the cell membrane contributes to whole cell Cl^- conductance and whether the amount of membrane-resident CFTR is enhanced during stimulation of the cells remain controversial. In the latter case, CFTR may accumulate in the cell membrane by enhanced exocytosis [98, 117–125] or by limited endocytosis [98, 99] which is controlled by amino acid stretches at both the N and C termini, which serve as internalization signals [92]. Exocytosis of CFTR was observed in *Xenopus* oocytes by 1 group [118, 125] and these data were confirmed in our laboratory by expressing CFTR in oocytes; CFTR was detected on the surface of *Xenopus* oocytes using an antibody that binds to an extracellular domain [126] and a FITS-labelled secondary antibody. The fluorescence signal was largely enhanced by stimula-

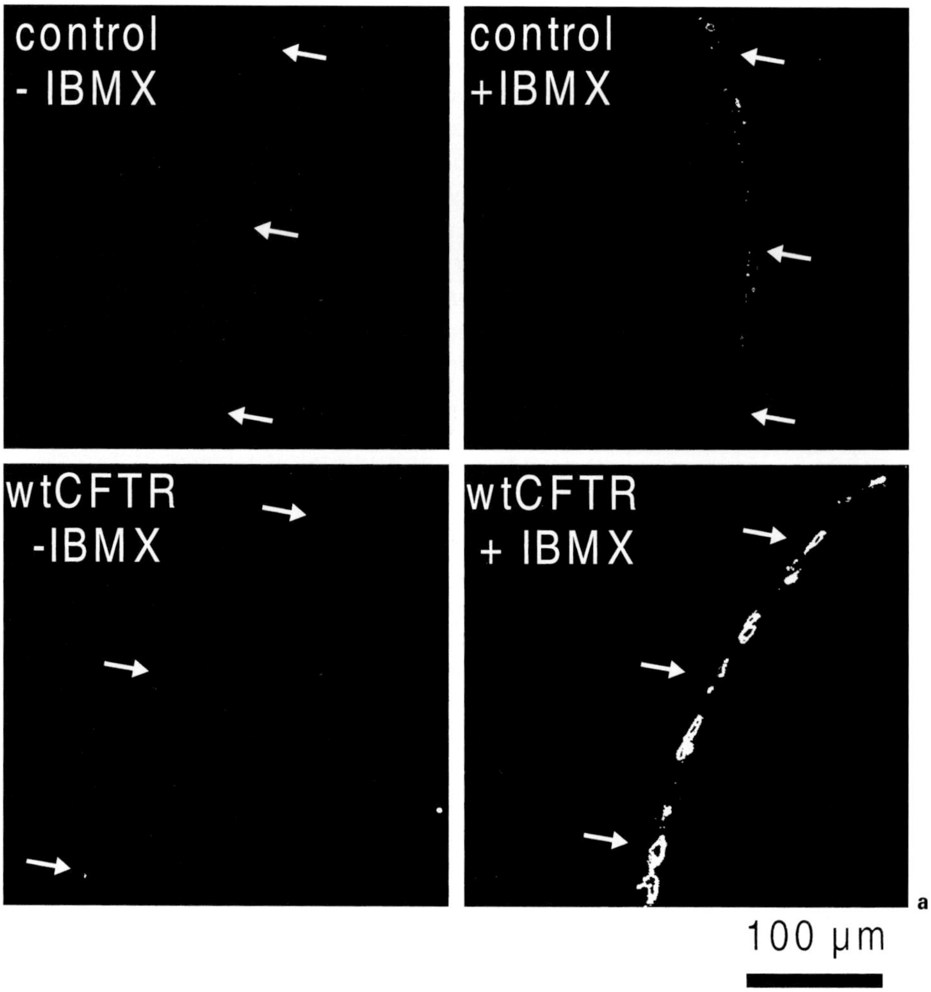

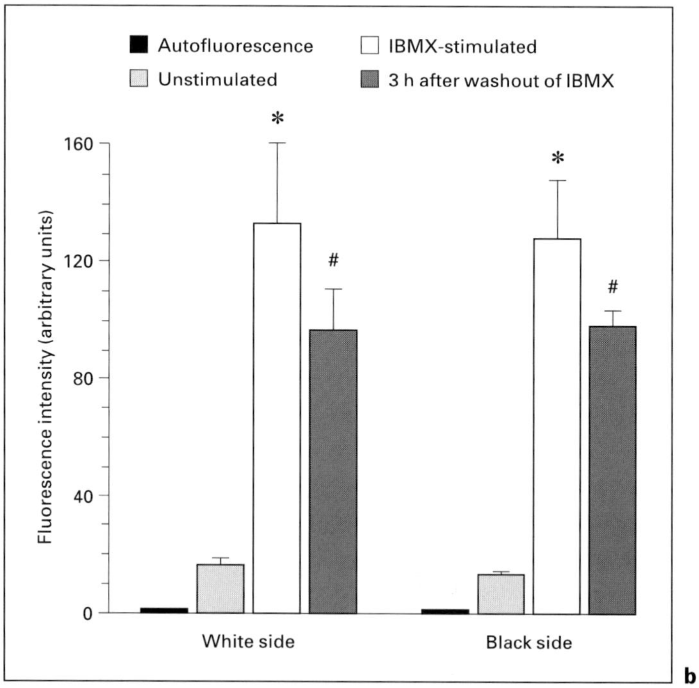

Fig. 1. Fluorescence labelling of CFTR in living oocytes using an antibody that binds to an extracellular loop of CFTR and a secondary FITS-labelled antibody. **a** Confocal microscopic images of the cell membrane that were taken before and after stimulation of oocytes expressing wtCFTR and water-injected control oocytes with 3-isobutyl-1-methylxanthine (IBMX). **b** Summary of the fluorescence signals measured in oocytes expressing wtCFTR. Fluorescence was detected separately on the white and black side of the oocytes. The autofluorescence of the oocyte membrane was small compared to specific labelling. However, fluorescence detected after incubation with antibodies was significantly increased when oocytes were stimulated with IBMX. A significant decrease in fluorescence intensity was observed 3 h after washout of the antibodies. Experiments (n = 5) were performed in paired fashion and oocytes were restained after every experimental step. *,# Significant difference: paired t test, $p < 0.05$.

tion of the oocytes (fig. 1). In contrast, only a weak fluorescence signal was detected when ΔF508-CFTR was expressed, and fluorescence was not increased further by cAMP-dependent stimulation (data not shown). This process of exocytosis of wtCFTR might be controlled by Gi proteins [127–129]. Despite all these results indicating exocytosis of CFTR, it has been shown that CFTR can be activated directly by PKA and ATP in isolated membrane patches [130]. In addition, other studies could not detect any significant changes in membrane capacitance and membrane surface area during stimulation of CFTR in Chinese hamster ovary cells and Calu-3 cells [124, 131].

Effects of CFTR on Exocytosis

In addition to the issue of CFTR trafficking the luminal membrane upon cAMP-dependent stimulation, there is substantial evidence that CFTR actually regulates the exocytosis of other proteins. For example, defects in mucin secretion in CF were reported several years ago [132, 133]. These reports indicated that constitutive, cAMP- and ATP-dependent mucin secretion was defective in CF pancreatic and airway epithelia [28, 134–136] but not in gallbladder epithelial cells [137]. Mucin is secreted by mucous cells of the trachea, pancreas, gallbladder and intestinal tract which all express CFTR [137–139], and it seems likely that both defective regulation of mucin secretion and enhanced expression of certain mucins contribute to alterations of mucous properties described in CF.

More recent studies have indicated a CFTR-dependent control of membrane exo- and endocytosis, and of the regulation of membrane turnover [102, 117, 118, 140, 141]. Further evidence came from studies demonstrating physical interaction of CFTR with syntaxin 1A, a protein known to be involved in neurosecretion and anchoring of CFTR to the cytoskeleton [142, 143]. The effects of CFTR on membrane turnover seem to be limited to clathrin-dependent endocytosis, i.e. endocytotic retrieval of a certain subfraction of vesicles [98, 140], while fluid phase endocytosis is not affected [144].

The regulatory effects of CFTR on other membrane conductances may be caused by CFTR-dependent regulation of exocytosis or endocytosis of ion channels. Accordingly, it was shown that ENaCs undergo clathrin-mediated endocytosis [145]. CFTR would be able to regulate in parallel the activity of several individual membrane conductances [3] through exocytosis and endocytosis. Depolymerization of the cytoskeleton during exocytosis and generation of short actin filaments might be another way in which CFTR interacts with other membrane proteins [121, 146–148]. Indeed such a modulatory action of short actin filaments has already been postulated [149, 150] for the CFTR-dependent regulation of ENaCs.

Taken together these results suggest that it is likely that regulation of CFTR by exocytosis is present in polarized epithelial cells and *Xenopus* oocytes, but not in other cell types [101]. Similar conflicting results had been anticipated when insulin-dependent insertion of the glucose transporter, GLUT4, was studied in cells other than adipocytes and myocytes, because the machinery for GLUT4 trafficking is only expressed in the native cells [101]. Given CFTR's multiple effects on other membrane conductances [3] and on mucin secretion, it is quite likely that CFTR acts as a regulator of vesicle traffic and protein exocytosis.

Pharmacological Interference with Maturation of CFTR

CF is caused by a variety of mutations that affect different functions of CFTR [22, 23]. Many of these mutant forms of CFTR, including the most common CFTR mutation, ΔF508, have residual Cl^- channel function and can be activated by PKA or alternative pathways. For ΔF508-CFTR, Cl^- conductance largely depends on the amount of protein that reaches maturation during biosynthesis [60, 61, 151]. There are several potential mechanisms by which residual Cl^- conductance, generated by mutant CFTR, can be enhanced. This includes: (i) increased expression of mutant CFTR; (ii) inhibition of CFTR degradation; (iii) activation of trafficking of mutant CFTR to the cell membrane; (iv) augmentation of intracellular stimulatory pathways, or (v) direct activation of mutant CFTR. All of these mechanisms represent potential pharmacological targets for the treatment of CF. Recent reviews offer a more detailed insight into CFTR's pharmacology [1, 25, 152]. Maturation of mutant CFTR protein during biosynthesis can be enhanced by several factors [60, 70, 153]. Glycerol and trimethylamine-N-oxide are known to stabilize conformation of both wtCFTR and ΔF508-CFTR by acting as chemical chaperons, thereby increasing the trafficking to the cell membrane and thus the amounts of CFTR present in the cell membrane. Unfortunately, very high concentrations of these chaperons are required to induce the effects on CFTR maturation [153, 154]. Recently, the compound deoxyspergualin was demonstrated to partially restore Cl^- conductance in cells overexpressing ΔF508-CFTR [155]. Deoxyspergualin is known to bind the heat shock proteins Hsp70 and Hsp90. Geldanamycin is another potential antichaperone drug [95, 156]. In contrast, inhibition of intracellular proteases or interference with ubiquitination is unlikely to restore Cl^- conductance because both would lead to accumulation of highly aggregated CFTR [47]. Thus, future studies will have to concentrate on the development of antichaperones. These might be good candidates for a future therapy of CF, particularly when administered in concert with compounds that activate mutant CFTR directly [157].

Summary and Conclusion

CFTR is an integral membrane protein with multiple cellular functions. It is characterized by complex regulation of both maturation and intracellular processing, and is activated by protein phosphorylation and protein kinase independent mechanisms. The tight control of processing makes the synthesis of CFTR a rather ineffective and temperature-sensitive process, during which 75% of wtCFTR and almost 100% of ΔF508-CFTR is degraded before leaving the ER. Several chaperons and proteolytic systems, like calnexin, Hsp70, Hsp90 and proteasomes participate in this degradation process. These chaperons could serve as new pharmacological targets for the treatment of CF, because CF is characterized by a lack of expression of functional CFTR. When applied in combination with compounds that boost CFTR activity independent of intracellular cAMP, these modulators of chaperons may be particularly useful.

CFTR trafficking to the cell membrane is currently not completely understood. In polarized epithelial cells and also *Xenopus* ooyctes, activation of Cl^- conductance occurs not only through PKA (and probably PKC)-dependent phosphorylation of CFTR that is already located in the cell membrane, but also via exocytosis of CFTR protein. This process is linked to an increase in membrane capacitance and is also detected by using fluorescence-labelled CFTR. In non-epithelial cells which overexpress CFTR, exocytosis is not a requirement for activation of CFTR and cAMP-dependent Cl^- conductance. Further studies have to concentrate on identifying the underlying mechanisms and additional proteins that participate in exocytosis. These proteins could be expressed differentially in different cells types. It will also be interesting to learn to what extent exocytosis of CFTR is connected to the defects of other membrane conductances and the related pathophysiological phenomena observed in CF.

Acknowledgments

We gratefully thank Prof. Dr. R. Greger for his support and helpful suggestions.

References

1 Kunzelmann K: The cystic fibrosis transmembrane conductance regulator and its function in epithelial transport. Rev Physiol Biochem Pharmacol 1999;137:1–70.
2 Pilewski JM, Frizzell RA: Role of CFTR in airway disease. Physiol Rev. 1999;79(suppl):215–255.
3 Kunzelmann K, Schreiber R: CFTR, a regulator of channels. J Membr Biol 1999;168:1–8.
4 Schwiebert EM, Benos DJ, Egan ME, Stutts MJ, Guggino WB: CFTR is a conductance regulator as well as a chloride channel. Physiol Rev. 1999;79(suppl):145–166.
5 Riordan JR, Rommels JM, Kerem B-S, Alon N, Rozmahel R, Grzelczak Z, Zielenski J, Plavsic SLN, Chou J-L, Drumm ML, Iannuzzi CM, Collins FS, Tsui L-C: Identification of the cystic fibrosis gene: Cloning and characterization of complementary DNA. Science 1989;245: 1066–1072.
6 Lukacs GL, Abdalla M, Kartner N, Riordan JR, Grinstein S: Conformational maturation of CFTR but its mutant counterpart (deltaF508) occurs in the endoplasmic reticulum and requires ATP. EMBO J. 1994;13:6076–6086.
7 Riordan JR: The cystic fibrosis transmembrane conductance regulator. Annu Rev Physiol 1993;55:609–630.
8 Mathews CJ, Tabcharani JA, Chang XB, Jensen TJ, Riordan JR, Hanrahan JW: Dibasic protein kinase A sites regulate bursting rate and nucleotide sensitivity of the cystic fibrosis transmembrane conductance regulator chloride channel. J Physiol (Lond) 1998;508:365–377.
9 Chang X-B, Tabcharani JA, Hou Y-X, Jensen TJ, Kartner N, Alon N, Hanrahan JW, Riordan JR: Protein kinase A (PKA) still activates CFTR chloride channels after mutagenesis of all 10 PKA consensus phosphorylation sites. J Biol Chem 1993;268:11304–11311.
10 Steagall WK, Kelley TJ, Marsick RJ, Drumm ML: Type II protein kinase A regulates CFTR in airway, pancreatic, and intestinal cells. Am J Physiol 1998;274:C819–C826.
11 Fuller CM, Benos DJ: CFTR! Am J Physiol 1992;263:C267–C286.
12 Chang X-B, Hou YX, Jensen TJ, Riordan JR: Mapping of cystic fibrosis transmembrane conductance regulator membrane topology by glycosylation site insertion. J Biol Chem 1994; 269:18572–18575.
13 Akabas MH, Kaufmann C, Cook TA, Archdeacon P: Amino acid residues lining the chloride channel of cystic fibrosis transmembrane conductance regulator. J Biol Chem 1995;269: 14865–14868.
14 Ko YH, Delannoy M, Pedersen PL: Cystic fibrosis transmembrane conductance regulator: The first nucleotide binding fold targets the membrane with retention of its ATP binding function. Biochemistry 1997;36:5053–5064.
15 Arispe N, Rojas E, Hartman J, Sorscher EJ, Pollard HB: Intrinsic anion channel activity of the recombinant first nucleotide binding fold domain of the cystic fibrosis transmembrane regulator protein. Proc Natl Acad Sci USA 1992;89:1539–1543.

16 Ko YH, Thomas PJ, Delannoy MR, Pedersen PL: The cystic fibrosis transmembrane conductance regulator. Overexpression, purification, and characterization of wild type and delta F508 mutant forms of the first nucleotide binding fold in fusion with the maltose-binding protein. J Biol Chem 1993;268:24330–24338.
17 Gruis DB, Price EM: The nucleotide binding folds of the cystic fibrosis transmembrane conductance regulator are extracellularly accessible. Biochemistry 1997;36:7739–7745.
18 Annereau JP, Wulbrand U, Vankeerberghen A, Cuppens H, Bontems F, Tummler B, Cassiman JJ, Stoven V: A novel model for the first nucleotide binding domain of the cystic fibrosis transmembrane conductance regulator. FEBS Lett 1997;407:303–308.
19 Bianchet MA, Ko YH, Amzel LM, Pedersen PL: Modeling of nucleotide binding domains of ABC transporter proteins based on a F1-ATPase/recA topology: Structural model of the nucleotide binding domains of the cystic fibrosis transmembrane conductance regulator (CFTR). J Bioenerg Biomembr 1997;29:503–524.
20 Hung LW, Wang IX, Nikaido K, Liu PQ, Ames GF, Kim SH: Crystal structure of the ATP-binding subunit of an ABC transporter. Nature 1998;396:703–707.
21 Sheppard DN, Rich DP, Ostedgaard LS, Gregory RJ, Smith AE, Welsh MJ: Mutations in CFTR associated with mild-disease-form Cl^- channels with altered pore properties. Nature 1993;362:160–164.
22 Welsh MJ, Smith AE: Molecular mechanisms of CFTR chloride channel dysfunction in cystic fibrosis. Cell 1993;73:1251–1254.
23 Tsui L-C, Durie P: Genotype and phenotype in cystic fibrosis. Hosp Pract 1997;32:115–118, 123–129, 134.
24 Wilschanski MA, Zielenski J, Markiewicz D, Tsui L-C, Corey M, Levison H, Durie PR: Correlation of sweat chloride concentration with classes of the cystic fibrosis transmembrane conductance regulator gene mutations. J Pediatr 1995;127:705–710.
25 Schultz BD, Singh AK, Devor DC, Bridges RJ: Pharmacology of CFTR chloride channel activity. Physiol Rev 1999;79(suppl):109–144.
26 Smith JJ, Travis SM, Greenberg EP, Welsh MJ: Cystic fibrosis airway epithelia fail to kill bacteria because of abnormal airway surface fluid. Cell 1996;85:229–236.
27 Barasch J, Kiss B, Prince A, Saiman L, Gruenert DC, al-Awqati Q: Defective acidification of intracellular organelles in cystic fibrosis. Nature 1991;352:70–73.
28 Kuver R, Ramesh N, Lau S, Savard C, Lee SP, Osborne WR: Constitutive mucin secretion linked to CFTR expression. Biochem Biophys Res Commun 1994;203:1457–1462.
29 Schreiber R, Greger R, Kunzelmann K: Cystic fibrosis transmembrane conductance regulator activates water conductance in *Xenopus* oocytes. Pflügers Arch 1997;434:841–847.
30 Schreiber R, Nitschke R, Greger R, Kunzelmann K: CFTR activates AQP3 in airway epithelial cells. J Biol Chem 1998;274:11811–11816.
31 Zeng W, Lee MG, Yan M, Diaz J, Benjamin I, Marino CR, Kopito RR, Freedman S, Cotton C, Muallem S, Thomas P: Immuno and functional characterization of CFTR in submandibular and pancreatic acinar and duct cells. Am J Physiol 1997;273:C442–C455.
32 Crawford I, Maloney PC, Zeitlin PL, Guggino WB, Hyde SC, Turley H, Gatter KC, Harris A, Higgins CF: Immunocytochemical localization of the cystic fibrosis gene product CFTR. Proc Natl Acad Sci USA 1991;88:9262–9266.
33 Reddy MM, Quinton PM: cAMP activation of CF-affected Cl^- conductance in both cell membranes of an absorptive epithelium. J Membr Biol 1992;130:49–62.
34 Reynaert I, Van Der Schueren B, Degeest G, Manin M, Cuppens H, Scholte B, Cassiman JJ: Morphological changes in the vas deferens and expression of the cystic fibrosis transmembrane conductance regulator (CFTR) in control, deltaF508 and knock-out CFTR mice during postnatal life. Mol Reprod Dev 2000;55(2): 125–135.
35 Gadsby DC, Nagel G, Hwang T-C: The CFTR chloride channel of mammalian heart. Annu Rev Physiol 1995;57:387–416.
36 Morales MM, Carroll TP, Morita T, Schwiebert EM, Devuyst O, Wilson PD, Lopes AG, Stanton BA, Dietz HC, Cutting GR, Guggino WB: Both the wild type and a functional isoform of CFTR are expressed in kidney. Am J Physiol 1996;270:F1038–F1048.
37 Mulberg AE, Resta LP, Wiedner EB, Altschuler SM, Jefferson DM, Broussard DL: Expression and localization of the cystic fibrosis transmembrane conductance regulator mRNA and its protein in rat brain. J Clin Invest 1995;96: 646–652.
38 Stanton BA: Cystic fibrosis transmembrane conductance regulator (CFTR) and renal function. Wien Klin Wochenschr 1997;109:457–464.
39 McDonald TV, Nghiem PT, Gardner P, Martens CL: Human lymphocytes transcribe the cystic fibrosis transmembrane conductance regulator gene and exhibit CF-defective cAMP-regulated chloride current. J Biol Chem 1992; 267:3242–3248.
40 Trapnell BC, Chu C-S, Paakko PK, Banks TC, Yoshimura K, Ferrans VJ, Chernick MS, Crystal RG: Expression of the cystic fibrosis transmembrane conductance regulator gene in the respiratory tract of normal individuals and individuals with cystic fibrosis. Proc Natl Acad Sci USA 1991;88:6565–6569.
41 Engelhardt JF, Zepeda M, Cohn JA, Yankaskas JR, Wilson JM: Expression of the cystic fibrosis gene in adult human lung. J Clin Invest 1994; 93:737–749.
42 Beck S, Kühr J, V.Schütz V, Röstermund T, Seydewitz H, Brandis M, Greger R, Kunzelmann K: No correlation of CF-phenotype, CFTR-Cl^- currents and expression of CFTR-mRNA. Pediatr Pulmonol 1999;27:251–259.
43 Sood R, Bear C, Auerbach W, Reyes E, Jensen T, Kartner N, Riordan JR, Buchwald M: Regulation of CFTR expression and function during differentiation of intestinal epithelial cells. EMBO J 1992;11:2487–2494.
44 Montrose-Rafizadeh C, Guggino WB, Montrose MH: Cellular differentiation regulates expression of Cl^- transport and cystic fibrosis transmembrane conductance regulator mRNA in human intestinal cells. J Biol Chem 1991; 266:4495–4499.
45 Montrose-Rafizadeh C, Blackmon DL, Hamosh A, Oliva MM, Hawkins AL, Curristin SM, Griffin CA, Yang VW, Guggino WB, Cutting GR: Regulation of cystic fibrosis transmembrane conductance regulator (CFTR) gene transcription and alternative RNA splicing in a model of developing intestinal epithelium. J Biol Chem 1992;267:19299–19305.
46 Cozens AL, Yezzi MJ, Kunzelmann K, Ohrui T, Chin L, Eng K, Finkbeiner E, Widdicombe JH, Gruenert DC: CFTR expression and chloride secretion in polarized immortal human bronchial epithelial cells. Am J Respir Cell Mol Biol 1994;10:38–47.
47 Kopito RR: Biosynthesis and degradation of CFTR. Physiol Rev 1999;79(suppl):167–173.
48 Mohammad-Panah R, Demolombe S, Riochet D, Leblais V, Loussouarn G, Pollard H, Baro I, Escande D: Hyperexpression of recombinant CFTR in heterologous cells alters its physiological properties. Am J Physiol 1998;274:C310–C318.
49 Matthews RP, McKnight GS: Characterization of the cAMP response element of the cystic fibrosis transmembrane conductance regulator gene promoter. J Biol Chem 1996;271:31869–31877.
50 Pittman N, Shue G, Leleiko NS, Walsh MJ: Transcription of cystic fibrosis transmembrane conductance regulator requires a CCAAT-like element for both basal and cAMP-mediated regulation. J Biol Chem 1995;270:28848–28857.
51 Breuer W, Kartner N, Riordan JR, Cabantchik ZI: Induction of expression of the cystic fibrosis transmembrane conductance regulator. J Biol Chem 1992;267:10465–10469.
52 Kunzelmann K, Allert N, Kubitz R, Breuer WV, Cabantchik ZI, Normann C, Schumann S, Leipziger J, Greger R: Forskolin- and PMA-pretreatment alter the acute electrical response of HT_{29} cells to cAMP, ATP, neurotensin, ionomycin and hypotonic cell swelling. Pflügers Arch 1994;428:76–83.
53 Breuer W, Glickstein H, Kartner N, Riordan JR, Ausiello DA, Cabantchik IZ: Protein kinase C mediates down-regulation of cystic fibrosis transmembrane conductance regulator levels in epithelial cells. J Biol Chem 1993;268: 13935–13939.
54 Bargon J, Trapnell BC, Yoshimura K, Dalemans W, Pavirani A, Lecocq JP, Crystak RG: Expression of the cystic fibrosis transmembrane conductance regulator gene can be regulated by protein kinase C. J Biol Chem 1992; 267:16056–16060.
55 Kang-ParksS, Dray-Charier N, Munier A, BbrahimiHorn C, Veissiere D, Picard J, Capeau J, Cherqui G, Lascols O: Role for PKC alpha and PKC epsilon in down-regulation of CFTR mRNA in a human epithelial liver cell line. J Hepatol 1998;28:250–262.

56 Bargon J, Trapnell BC, Chu CS, Rosenthal ER, Yoshimura K, Guggino WB, Dalemans W, Pavirani A, Lecocq JP, Crystal RG: Down-regulation of cystic fibrosis transmembrane conductance regulator gene expression by agents that modulate intracellular divalent cations. Mol Cell Biol 1992;12:1872–1878.
57 Breuer W, Slotki IN, Ausiello DA, Cabantchik IZ: Induction of multidrug resistance downregulates the expression of CFTR in colon epithelial cells. Am J Physiol 1993;265:C1711–C1715.
58 Kunzelmann K, Slotki IN, Klein P, Koslowsly T, Ausiello DA, Greger R, Cabantchik ZI: Effects of P-glycoprotein expression on cyclic AMP and volume activated ion fluxes and conductances in HT_{29} colon carcinoma cells. J Cell Physiol 1994;161:393–406.
59 Kunzelmann K, Kathöfer S, Hipper A, Gruenert DC, Greger R: Culture-dependent expression of Na^+ conductances in airway epithelial cells. Pflügers Arch 1996;431:578–586.
60 Jensen TJ, Loo MA, Pind S, Williams DB, Goldberg AL, Riordan JR: Multiple proteolytic systems, including the proteasome, contribute to CFTR processing. Cell 1995;83:13–20.
61 Ward CL, Omura S, Kopito RR: Degradation of CFTR by the ubiquitin-proteasome pathway. Cell 1995;83:121–127.
62 Bebok Z, Mazzochi C, King SA, Hong JS, Sorscher EJ: The mechanism underlying cystic fibrosis transmembrane conductance regulator transport from the endoplasmic reticulum to the proteasome includes Sec61beta and a cytosolic, deglycosylated intermediary. J Biol Chem 1998;273:29873–29878.
63 Christian WW, Fabunmi RP, Lee MG, Marino CR, Muallem S, Demartino GN, Thomas PJ: Dynamic Association of Proteasomal Machinery with the Centrosome. J Cell Biol 1999;145: 481–490.
64 Johnston JA, Ward CL, Kopito RR: Aggresomes: A cellular response to misfolded proteins. J Cell Biol 1998;143:1883–1898.
65 Marshall J, Fang S, Ostedgaard LS, O'Riordan CR, Ferrara D, Amara JF, Hoppe H, Scheule RK, Welsh MJ, Smith AE: Stoichiometry of recombinant cystic fibrosis transmembrane conductance regulator in epithelial cells and its functional reconstitution into cells in vitro. J Biol Chem 1994;269:2987–2995.
66 Zerhusen B, Zhao J, Xie J, Davis PB, Ma J: A single conductance pore for chloride ions formed by two cystic fibrosis transmembrane conductance regulator molecules. J Biol Chem 1999;274:7627–7630.
67 Ward CL, Kopito RR: Intracellular turnover of cystic fibrosis transmembrane conductance regulator. Inefficient processing and rapid degradation of wild-type and mutant proteins. J Biol Chem 1994;269:25710–25718.
68 Morris AP, Cunningham SA, Benos DJ, Frizzell RA: Glycosylation status of endogenous CFTR does not affect cAMP-stimulated Cl^- secretion in epithelial cells. Am J Physiol 1993; 265:C688–C694.
69 Jiang X, Hill WG, Pilewski JM, Weisz OA: Glycosylation differences between a cystic fibrosis and rescued airway cell line are not CFTR dependent. Am J Physiol 1997;273: L913–L920.
70 Denning GM, Anderson MP, Amara JF, Marshall J, Smith AE, Welsh MJ: Processing of mutant cystic fibrosis transmembrane conductance regulator is temperature-sensitive. Nature 1992;358:761–764.
71 French PJ, Van Doorninck JH, Peters RH, VerbeekeE, Ameen NA, Marino CR, de Jonge HR, Bijman J, Scholte BJ: A delta F508 mutation in mouse cystic fibrosis transmembrane conductance regulator results in a temperature-sensitive processing defect in vivo. J Clin Invest 1996;98:1304–1312.
72 Cunningham SA, Worrell RT, Benos DJ, Frizzell RA: cAMP-stimulated ion currents in xenopus oocytes expressing CFTR cRNA. Am J Physiol 1992;262:C783–C788.
73 Briel M, Greger R, Kunzelmann K: Cl^- transport by CFTR contributes to the inhibition of epithelial Na^+ channels in *Xenopus* ooyctes coexpressing CFTR and ENaC. J Physiol (Lond) 1998;508.3:825–836.
74 Cheng SH, Fang SL, Zabner J, Marshall J, Piraino S, Schiavi SC, Jefferson DM, Welsh MJ, Smith AE: Functional activation of the cystic fibrosis trafficking mutant delta F508-CFTR by overexpression. Am J Physiol 1995; 268:L615–L624.
75 Qu BH, Strickland EH, Thomas PJ: Localization and suppression of a kinetic defect in cystic fibrosis transmembrane conductance regulator folding. J Biol Chem 1997;272:15739–15744.
76 Denning GM, Ostedgaard LS, Welsh MJ: Abnormal localization of cystic fibrosis transmembrane conductance regulator in primary cultures of cystic fibrosis airway epithelia. J Cell Biol 1992;118:551–559.
77 Puchelle E, Gaillard D, Ploton D, Hinnrasky J, Fuchey C, Boutterin MC, Jacquot J, Dreyer D, Pavirani A, Dalemans W: Differential localization of the cystic fibrosis transmembrane conductance regulator in normal and cystic fibrosis airway epithelium. Am J Respir Cell Mol Biol 1992;7:485–491.
78 Ostergaard LS, Zeiher B, Welsh MJ: Processing of CFTR bearing the P574H mutation differs from wild-type and (&Dgr;)F508-CFTR. J Cell Sci 1999;112:2091–2098.
79 Seibert FS, Linsdell P, Loo TW, Hanrahan JW, Clarke DM, Riordan JR: Disease-associated mutations in the fourth cytoplasmic loop of cystic fibrosis transmembrane conductance regulator compromise biosynthetic processing and chloride channel activity. J Biol Chem 1996;271:15139–15145.
80 Seibert FS, Jia Y, Mathews CJ, Hanrahan JW, Riordan JR, Loo TW, Clarke DM: Disease-associated mutations in cytoplasmic loops 1 and 2 of cystic fibrosis transmembrane conductance regulator impede processing or opening of the channel. Biochemistry 1997;36:11966–11974.
81 Gregory RJ, Rich DP, Cheng SH, Souza DW, Paul S, Manavalan P, Anderson MP, WwelshMJ, Smith AE: Maturation and function of cystic fibrosis transmembrane conductance regulator variants bearing mutations in putative nucleotide-binding domains 1 and 2. Mol Cell Biol 1991;11:3886–3893.
82 Sheppard DN, Travis SM, Ishihara H, Welsh MJ: Contribution of proline residues in the membrane-spanning domains of cystic fibrosis transmembrane conductance regulator to chloride channel function. J Biol Chem 1996;271: 14995–15001.
83 Katzmann DJ, Epping EA, Moye-Rowley WS: Mutational disruption of plasma membrane trafficking of *Saccharomyces cerevisiae* Yor1p, a homologue of mammalian multidrug resistance protein. Mol Cell Biol 1999;19:2998–3009.
84 Teem JL, Berger HA, Ostedgaard LS, Rich DP, Tsui LC, Welsh MJ: Identification of revertants for the cystic fibrosis delta F508 mutation using STE6-CFTR chimeras in yeast. Cell 1993;73:335–346.
85 Dalemans W, Barbry P, Champigny G, Jallat S, Dott K, Dreyer D, Crystal RG, Pavirani A, Lecocq J-P, Lazdunski M: Altered chloride ion channel kinetics associated with the deltaF508 cystic fibrosis mutation. Nature 1992;354: 526–528.
86 Li C, Ramjeesingh M, Reyes E, Jensen TJ, Chang X-B, Rommens JA, Bear CE: The cystic fibrosis mutation (deltaF508) does not influence the chloride channel activity of CFTR. Nat Genet 1993;3:311–316.
87 Lukacs GL, Chang XB, Bear C, Kartner N, Mohamed A, Riordan JR, Grinstein S: The delta F508 mutation decreases the stability of cystic fibrosis transmembrane conductance regulator in the plasma membrane. Determination of functional half-lives on transfected cells. J Biol Chem 1993;268:21592–21598.
88 Kartner N, Augustinas T, Jensen TJ, Naismith AL, Riordan JR: Mislocalization of deltaF508 CFTR in cystic fibrosis sweat gland. Nat Genet 1992;1:321–327.
89 Xiong X, Chong E, Skach WR: Evidence that endoplasmic reticulum (ER)-associated degradation of cystic fibrosis transmembrane conductance regulator is linked to retrograde translocation from the ER membrane. J Biol Chem 1999;274:2616–2624.
90 Pasyk EA, Foskett JK: Mutant (delta F508) cystic fibrosis transmembrane conductance regulator Cl^- channel is functional when retained in endoplasmic reticulum of mammalian cells. J Biol Chem 1995;270:12347–12350.
91 Sommer T, Wolf DH: Endoplasmic reticulum degradation: Reverse protein flow of no return. FASEB J 1997;11:1227–1233.
92 Pind S, Riordan JR, Williams DB: Participation of the endoplasmic reticulum chaperone calnexin (p88, IP90) in the biogenesis of the cystic fibrosis transmembrane conductance regulator. J Biol Chem 1994;269:12784–12788.

93 Yang Y, Janich S, Cohn JA, Wilson JM: The common variant of cystic fibrosis transmembrane conductance regulator is recognized by hsp70 and degraded in a pre-Golgi nonlysosomal compartment. Proc Natl Acad Sci USA 1993;90:9480–9484.
94 Meacham GC, Lu Z, King S, Sorscher E, Tousson A, Cyr DM: The Hdj-2/Hsc70 chaperone pair facilitates early steps in CFTR biogenesis. EMBO J 1999;18:1492–1505.
95 Loo MA, Jensen TJ, Cui L, Hou Y, Chang XB, Riordan JR: Perturbation of Hsp90 interaction with nascent CFTR prevents its maturation and accelerates its degradation by the proteasome. EMBO J 1998;17:6879–6887.
96 Dupuit F, Kalin N, Brezillon S, Hinnrasky J, Tümmler B, Puchelle E: CFTR and differentiation markers expression in non-CF and delta F-508 homozygous CF nasal epithelium. J Clin Invest 1995;96:1601–1611.
97 Kalin N, Claass A, Sommer M, Puchelle E, Tümmler B: DeltaF508 CFTR protein expression in tissues from patients with cystic fibrosis [see comments]. J Clin Invest 1999; 103:1379–1389.
98 Lukacs GL, Segal G, Kartner N, Grinstein S, Zhang F: Constitutive internalization of cystic fibrosis transmembrane conductance regulator occurs via clathrin-dependent endocytosis and is regulated by protein phosphorylation. Biochem J 1997;328:353–361.
99 Bradbury NA, Cohn JA, Venglarik CJ, Bridges RJ: Biochemical and biophysical identification of cystic fibrosis transmembrane conductance regulator chloride channels as components of endocytic clathrin coated vesicle. J Biol Chem 1994;269:8296–8302.
100 Lukacs GL, Chang XB, Kartner N, Rotstein OD, Riordan JR, Grinstein S: The cystic fibrosis transmembrane regulator is present and functional in endosomes. Role as a determinant of endosomal pH. J Biol Chem 1992; 267:14568–14572.
101 Bradybury NA: Intracellular CFTR: Localization and function. Physiol Rev 1999;79 (suppl):175–191.
102 Biwersi J, Verkman AS: Functional CFTR in endosomal compartment of CFTR-expressing fibroblasts and T84 cells. Am J Physiol 1994;266:C149–C156.
103 Gluck D, Kelly S, al-Awqati Q: The proton translocating ATPase responsible for urinary acidification. J Biol Chem 1982;257:9230–9233.
104 Root KV, Engelhardt JF, Post M, Wilson JW, van Dyke RW: CFTR does not alter acidification of L cell endosomes. Biochem Biophys Res Commun 1994;205:396–401.
105 Seksek O, Biwersi J, Verkman AS: Evidence against defective trans-Golgi acidification in cystic fibrosis. J Biol Chem 1996;271:15542–15548.
106 Imundo L, Barasch J, Prince A, al-Awqati Q: Cystic fibrosis epithelial cells have a receptor for pathogenic bacteria on their apical surface (Published erratum appears in Proc Natl Acad Sci USA 1995;92:11322.) Proc Natl Acad Sci USA 1995;92:3019–3023.
107 Hill WG, Harper GS, Rozaklis T, Boucher RC, Hopwood JJ: Organ-specific over-sulfation of glycosaminoglycans and altered extracellular matrix in a mouse model of cystic fibrosis. Biochem Mol Med 1997;62:113–122.
108 DosanJH A, Lencer W, Brown D, Ausiello DA, Stow JL: Heterologous expression of deltaF508 CFTR results in decreased sialylation of membrane glycoconjugates. Am J Physiol 1994;266:C360–C366.
109 Cheng PW, Boat TF, Cranfill K, Yankaskas JR, Boucher RC: Increased sulfation of glycoconjugates by cultured nasal epithelial cells from patients with cystic fibrosis. J Clin Invest 1989;84:68–72.
110 Poulsen JH, Machen TE: HCO_3^--dependent pH_i regulation in tracheal epithelial cells. Pflügers Arch 1996;432:546–554.
111 Poulsen JH, Fischer H, Illek B, Machen TE: Bicarbonate conductance and pH regulatory capability of cystic fibrosis transmembrane conductance regulator. Proc Natl Acad Sci USA 1994;91:5340–5344.
112 Gottlieb RA, Dosanjh A: Mutant cystic fibrosis transmembrane conductance regulator inhibits its acidification and apoptosis in C127 cells: Possible relevance to cystic fibrosis. Proc Natl Acad Sci USA 1996;93:3587–3591.
113 Wei LY, Hoffman MM, Roepe PD: Altered pHi regulation in 3T3/CFTR clones and their chemotherapeutic drug-selected derivatives. Am J Physiol 1997;272:C1642–C1653.
114 Davies JC, Stern M, Dewar A, Caplen NJ, Munkonge FM, Pitt T, Sorgi F, Huang L, Bush A, Geddes DM, Altin EW: CFTR gene transfer reduces the binding of *Pseudomonas aeruginosa* to cystic fibrosis respiratory epithelium. Am J Respir Cell Mol Biol 1997;16: 657–663.
115 Pier GB, Grout M, Zaidi TS: Cystic fibrosis transmembrane conductance regulator is an epithelial cell receptor for clearance of *Pseudomonas aeruginosa* from the lung. Proc Natl Acad Sci USA 1997;94:12088–12093.
116 Pier GB, Grout M, Zaidi T, Meluleni G, Mueschenborn SS, Banting G, Ratcliff R, Evans MJ, Colledge WH: *Salmonella typhi* uses CFTR to enter intestinal epithelial cells. Nature 1998;393:79–82.
117 Bradbury NA, Jilling T, Gabor B, Sorscher EJ, Bridges RJ, Kirk KL: Regulation of plasma membrane recycling by CFTR. Science 1992;256:530–531.
118 Takahasi A, Watkins SC, Howard M, Frizzell RA: CFTR-dependent membrane insertion is linked to stimulation of the CFTR Cl^- channel. Am J Physiol 1996;271:C1887–C1894.
119 Ameen NA, Martensson B, Bourguinon L, Marino C, Isenberg J, McLaughlin GE: CFTR channel insertion to the apical surface in rat duodenal villus epithelial cells is upregulated by VIP in vivo. J Cell Sci 1999;112: 887–894.
120 Prince LS, Workman RB, Marchase RB: Rapid endocytosis of the cystic fibrosis transmembrane conductance regulator chloride channel. Proc Natl Acad Sci USA 1994;91: 5192–5196.
121 Shapiro M, Matthews J, Hecht G, Delp C, Madara JL: Stabilization of F-actin prevents cAMP-elicited Cl^- secretion in T84 cells. J Clin Invest 1991;87:1903–1909.
122 Lehrich RW, Aller SG, Webster P, Marino CR, Forrest JN: Vasoactive intestinal peptide, forskolin, and genistein increase apical CFTR trafficking in the rectal gland of the spiny dogfish, *Squalus acanthias*. Acute regulation of CFTR trafficking in an intact epithelium. J Clin Invest 1998;101:737–745.
123 McNicholas CM, Nason MW, Jr., Guggino WB, Schwiebert EM, Hebert SC, Giebisch G, Egan ME: A functional CFTR-NBF1 is required for ROMK2-CFTR interaction. Am J Physiol 1997;273:F843–F848.
124 Hug MJ, Thiele IE, Greger R: The role of exocytosis in the activation of the chloride conductance in Chinese hamster ovary cells (CHO) stably expressing CFTR. Pflügers Arch 1997;434:779–784.
125 Howard M, Jilling T, Duvall M, Frizzell RA: cAMP-regulated trafficking of epitope-tagged CFTR. Kidney Int 1996;49:1642–1648.
126 Demolombe S, Baro I, Laurent M, Hongre AS, Pavorani A, Escande D: Abnormal subcellular localization of mutated CFTR protein in a cystic fibrosis epithelial cell line. Eur J Cell Biol 1994;65:214–219.
127 Schwiebert EM, Gruenert DC, Guggino WB, Stanton BA: G protein G alpha i-2 inhibits outwardly rectifying chloride channels in human airway epithelial cells. Am J Physiol 1995;269:C451–C456.
128 Schwiebert EM, Gesek F, Ercolani L, Wjasow C, Gruenert DC, Karlson K, Stanton BA: Heterotrimeric G proteins, vesicle trafficking, and CFTR Cl^- channels. Am J Physiol 1994;267:C272–C281.
129 Schwiebert EM, Kizer N, Gruenert DC, Stanton BA: GTP-binding proteins inhibit cAMP activation of chloride channels in cystic fibrosis airway epithelial cells. Proc Natl Acad Sci USA 1992;89:10623–10627.
130 Tabcharani JA, Chang X-B, Riordan JR, Hanrahan JW: Phosphorylation-regulated Cl^- channel in CHO cells stably expressing the cystic fibrosis gene. Nature 1991;352: 628–631.
131 Loffing J, Moyer BD, McCoy D, Stanton BA: Exocytosis is not involved in activation of Cl^- secretion via CFTR in Calu-3 airway epithelial cells [see comments]. Am J Physiol 1998; 275:C913–C920.
132 Blomfield J, Dascalu J, Lennep EW, Brown JM: Hypersecretion of zymogen granules in the pathogenesis of cystic fibrosis. Gut 1973; 14:558–565.
133 Frates RCJ, Kaizu TT, Last JA: Mucus glycoproteins secreted by respiratory epithelial tissue from cystic fibrosis patients. Pediatr Res 1983;17:30–34.
134 Montserrat C, Merten M, Figarella C: Defective ATP-dependent mucin secretion by cystic fibrosis pancreatic epithelial cells. FEBS Lett 1996;393:264–268.

135 Mergey M, Lemnaouar M, Veissiere D, Perricaudet M, Gruenert DC, Picard J, Capeau J, Brahimi-Horn MC, Paul A: CFTR gene transfer corrects defective glycoconjugate secretion in human CF epithelial tracheal cells. Am J Physiol 1995;269:L855–L864.

136 Pereira MM, Dormer RL, McPherson MA: Mucin secretion mediated by the cystic fibrosis gene protein, CFTR. Biochem Soc Trans 1995;23:532S.

137 Peters RH, French PJ, van Doorninck JH, Lamblin G, Ratcliff R, Evans MJ, Colledge WH, Bijman J, Scholte BJ: CFTR expression and mucin secretion in cultured mouse gallbladder epithelial cells. Am J Physiol 1996; 271:G1074–G1083.

138 Finkbeiner WE, Shen B-Q, Widdicombe JH: Chloride secretion and function of serous and mucous cells of human airway glands. Am J Physiol 1994;267:L206–L210.

139 Yamaya M, Finkbeiner WE, Widdicombe JH: Altered ion transport by tracheal glands in cystic fibrosis. Am J Physiol 1991;261: L491–L494.

140 Biwersi J, Emans N, Verkman AS: Cystic fibrosis transmembarne conductance regulator activation stimulates endosome fusion in vivo. Proc Natl Acad Sci USA 1996;93: 12484–12489.

141 Santos GF, Reenstra WW: Activation of the cystic fibrosis transmembrane regulator by cyclic AMP is not correlated with inhibition of endocytosis. Biochim Biophys Acta 1994; 1195:96–102.

142 Naren AP, Nelson DJ, Xie W, Jovov B, Pevsner J, Bennett MK, Benos DJ, Quick MW, Kirk KL: Regulation of CFTR chloride channels by syntaxin and Munc18 isoforms. Nature 1997;390:302–305.

143 Short DB, Trotter KW, Reczek D, Kreda SM, Bretscher A, Boucher RC, Stutts MJ, Milgram SL: An apical PDZ protein anchors the cystic fibrosis transmembrane conductance regulator to the cytoskeleton. J Biol Chem 1998;273:19797–19801.

144 Spilmont C, Hinnrasky J, Zahm JM, Jacquot J, Puchelle E, Kuhry JG: CFTR is involved in membrane endocytosis but not in fluid-phase and receptor-mediated endocytosis in human respiratory epithelial cells. Biochem Biophys Res Commun 1996;227:182–188.

145 Shimkets RA, Lifton RP, Canessa CM: The activity of the epithelial sodium channel is regulated by clathrin-mediated endocytosis. J Biol Chem 1997;272:25537–25541.

146 Hug T, Koslowsky T, Ecke D, Greger R, Kunzelmann K: Actin-dependent activation of ion conductances in bronchial epithelial cells. Pflügers Arch 1995;429:682–690.

147 Prat AG, Xiao Y-F, Ausiello DA, Cantiello HF: cAMP independent regulation of CFTR by the actin cytoskeleton. Am J Physiol 1995; 268:C1552–C1561.

148 Fischer H, Illek B, Machen TE: The actin filament disrupter cytochalasin D activates the recombinant cystic fibrosis transmembrane conductance regulator C^- channel in mouse 3T3 fibroblasts. J Physiol (Lond) 1995;489: 745–754.

149 Berdiev BK, Prat AG, Cantiello HF, Ausiello DA, Fuller CM, Jovov B, Benos DJ, Ismailov II: Regulation of epithelial sodium channels by short actin filaments. J Biol Chem 1996; 271:17704–17710.

150 Ismailov II, Berdiev BK, Shlyonsky VG, Fuller CM, Prat AG, Jovov B, Cantiello HF, Ausiello DA, Benos DJ: Role of actin in regulation of epithelial sodium channels by CFTR. Am J Physiol 1997;272:C1077–C1086.

151 Drumm ML, Wilkinson DJ, Smit LS, Worrell RT, Strong TV, Frizzell RA, Dawson DC, Collins FS: Chloride conductance expressed by deltaF508 and other mutant CFTRs in *Xenopus* oocytes. Science 1991;254:1797–1799.

152 Riordan JR: Therapeutic strategies for treatment of CF based on knowledge of CFTR. Pediatr Pulmonol Suppl 1999;18:83–87.

153 Sato S, Ward CL, Krouse ME, Wine JJ, Kopito RR: Glycerol reverses the misfolding phenotype of the most common cystic fibrosis mutation. J Biol Chem 1996;271:635–638.

154 Brown CR, Hong-Brown LQ, Biwersi J, Verkman AS, Welch WJ: Chemical chaperones correct the mutant phenotype of the delta F508 cystic fibrosis transmembrane conductance regulator protein. Cell Stress Chaperones 1996;1:117–125.

155 Jiang C, Fang SL, Xiao YF, O'Connor SP, Nadler SG, Lee DW, Jefferson DM, Kaplan JM, Smith AE, Cheng SH: Partial restoration of cAMP-stimulated CFTR chloride channel activity in DeltaF508 cells by deoxyspergualin. Am J Physiol 1998;275:C171–C178.

156 Roe SM, Prodromou C, O'Brien R, Ladbury JE, Piper PW, Pearl LH: Structural basis for inhibition of the Hsp90 molecular chaperone by the antitumor antibiotics radicicol and geldanamycin. J Med Chem 1999;42:260–266.

157 Loo TW, Clarke DM: Correction of defective protein kinesis of human P-glycoprotein mutants by substrates and modulators. J Biol Chem 1997;272:709–712.

158 Steagall WK, Drumm ML: Stimulation of cystic fibrosis transmembrane conductance regulator-dependent short-circuit currents across DeltaF508 murine intestines. Gastroenterology 1999;116:1379–1388.

Experimental Nephrology

Exp Nephrol 2000;8:343–350

Mechanisms of Renal Calcium Transport

Peter A. Friedman

Departments of Pharmacology and of Medicine, University of Pittsburgh School of Medicine, Pittsburgh, Pa., USA

Key Words
Calcium excretion · Nephron segments · Calcium channels · Ion channels · Parathyroid hormone · Diuretics · Gitelman's syndrome

Abstract
The kidneys play a key role in the integrated regulation of calcium homeostasis. Calcium absorption takes place throughout the nephron. Proximal tubules, thick ascending limbs of Henle's loop, and distal tubules are the major sites of calcium absorption. The mechanisms of absorption vary significantly from one segment to another, as does the extent of hormonal regulation. At one extreme is the considerable reabsorption by proximal tubules that proceeds primarily, if not entirely, by a paracellular pathway that is not regulated by hormones or drugs. In thick ascending limbs, calcium absorption occurs through a combination of transcellular and paracellular routes. The active, transcellular component is regulated by parathyroid hormone (PTH) and calcitonin, whereas the passive, paracellular route is governed by the extent of concomitant sodium absorption. At the other extreme is the distal tubule, where calcium absorption is entirely transcellular and is regulated by PTH, $1,25(OH)_2$vitamin D_3, calcitonin, and by calcium-sparing drugs such as thiazide-type diuretics. The present review focuses on recent insights into the mechanisms of transcellular calcium movement and highlights the discovery of an epithelial calcium channel, ECaC, that may mediate calcium entry in distal tubules.

Introduction

Extracellular calcium concentrations are tightly and coordinately regulated through the integrated participation of the parathyroid glands, kidneys, intestines, and bone, which is the repository for stored calcium. In response to even modest declines in serum ionized calcium, parathyroid gland cells secreted parathyroid hormone (PTH), which exerts both direct and indirect actions in defense of extracellular calcium; renal calcium excretion is reduced and osseous calcium is mobilized. Further bolstering these actions, PTH activates renal 1α,25-vitamin D hydroxylase, which converts $25(OH)D_3$ to its biologically active form, $1,25(OH)_2D_3$. $1,25(OH)_2D_3$, in turn, stimulates intestinal calcium and phosphate absorption and exerts a permissive, enhancing action on PTH-dependent renal calcium absorption. In this review, the cellular mechanisms by which calcium transport is accomplished and how they are regulated by hormones and drugs are described.

Calcium is absorbed throughout the nephron with the principal sites being proximal tubules, thick ascending

KARGER
Fax + 41 61 306 12 34
E-Mail karger@karger.ch
www.karger.com

Accessible online at:
www.karger.com/journals/exn

Peter A. Friedman, PhD
Department of Pharmacology, University of Pittsburgh School of Medicine
E-1347 Biomedical Science Tower
Pittsburgh, PA 15261 (USA)
Tel. +1 412 383 7783, Fax +1 412 648 1945, E-Mail paf10@pitt.edu

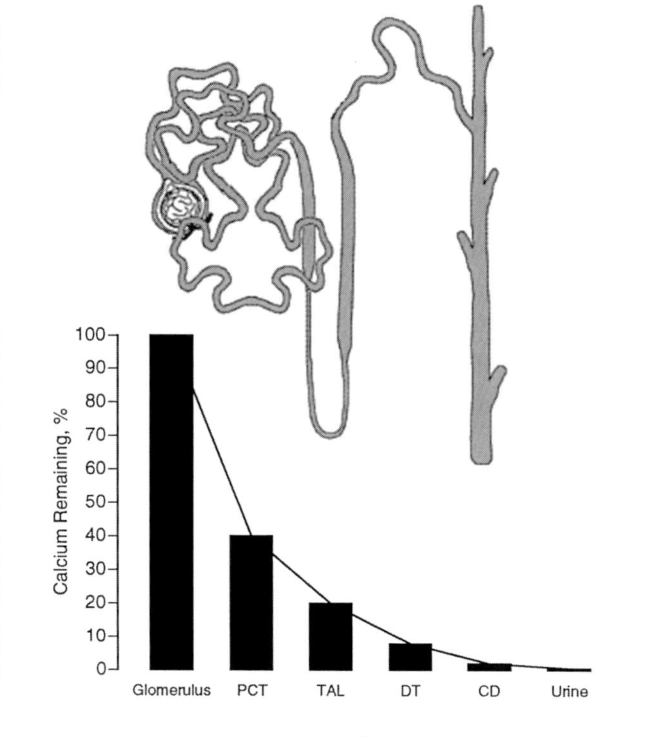

Fig. 1. Segmental absorption of calcium by key nephron segments. The graph displays the percentage of calcium remaining in the tubule at the end of the tubular segments designated below each bar. Since calcium is extensively bound to plasma proteins, absorption is normalized to that at the glomerulus. PCT = Proximal convoluted tubule; TAL = thick ascending limb of Henle's loop; DT = distal tubule; CD = collecting duct.

limbs, and distal tubules. However, the magnitude and mechanisms differ importantly in how this occurs between one segment and another. Likewise, the sites of hormonal and drug effects on calcium absorption are spatially separated. The majority of calcium is reabsorbed by proximal tubules (fig. 1). Progressively smaller amounts of calcium are recovered as the incipient urine passes through the nephron. Notably, however, it is on these smaller fractions of calcium transport that are reabsorbed by more distal nephron segments where the regulatory effects are achieved. From a simple perspective then, the bulk of calcium reabsorption occurs at the beginning of the nephron, where it is essentially unregulated. The tuning and fine adjustment of calcium recovery occurs at downstream sites in thick limbs and especially in distal tubules. In a healthy adult, generally $<1-2\%$ of the calcium initially filtered at the glomerulus is excreted in the voided urine.

Tubular Absorption of Calcium

Routes of Calcium Absorption

As noted above, proximal tubules absorb most of the filtered calcium. As this implies, not all the serum calcium is filtered by the glomerulus. The reason for this is that nearly half of the total serum calcium is bound to plasma proteins, which are not filtered. Only free calcium and calcium that is complexed to small anions pass through the slit membranes of the glomerulus and enter the tubular fluid of the proximal nephron. By the end of the proximal convoluted tubule, some 60–70% of the filtered calcium is absorbed.

Since the ratio of calcium in the tubular fluid to that in the glomerular ultrafiltrate hovers between 1.0 and 1.2, it is generally accepted that proximal calcium transport is essentially an isoosmotic process that is energetically passive. This suggests that calcium absorption in proximal tubules proceeds through the paracellular pathway between adjacent cells. Nonetheless, several detailed analyses, where driving forces for passive calcium movement were eliminated, adduced evidence that 10–15% of the reabsorption is active, implicating a cellular pathway for this process. Remarkably little is known about possible transcellular calcium absorption in proximal tubules.

Medullary and cortical thick ascending limbs reabsorb about 20% of the filtered calcium. In these segments calcium is transported by a combination of cellular and paracellular routes. Basal calcium absorption proceeds through the paracellular pathway, where its rate of movement is governed by the prevailing electrochemical driving forces, which, in turn, are established by the extent of sodium absorption. More will be said of this shortly. The cellular component of calcium absorption in the thick limbs is regulated by PTH and calcitonin. Thus, in contrast to proximal tubules, where nearly all of the transepithelial calcium absorption proceeds through the paracellular pathway, in medullary and cortical thick ascending limbs transport follows parallel absorption routes, a passive mechanism through the paracellular path and an active mechanism that is transcellular. A third permutation of this scheme occurs in distal tubules.

Distal tubules, consisting of distal convoluted tubules (DCTs) and connecting tubules, are responsible for reabsorbing 8–10% of the filtered load. In these segments, the calcium permeability of the tight junctions is low and passive calcium absorption is absent; all calcium absorption is active and follows a transcellular route.

Mechanisms of Calcium Absorption

Transepithelial absorption of calcium, like that of other solutes, can be conceptually described by the relation: calcium absorption = passive transport + active transport, where passive absorption is the sum of diffusion and solvent drag, and active transport is an energetically dependent, transcellular process. Cellular calcium absorption across polarized epithelial cells is a two-step process, where entry across apical (mucosal, luminal) membranes is followed by extrusion across basolateral (serosal, contraluminal) membranes into interstitial fluid and thence into the circulation. Calcium influx down its electrochemical gradient across apical membranes is generally thought to be mediated by calcium channels. Basolateral efflux, in contrast, involves energy-dependent extrusion that is accomplished by the plasma membrane Ca^{2+}-ATPase (PMCA) and the Na^+/Ca^{2+} exchanger.

Calcium absorption by proximal tubules is mediated by a combination of passive and active transport mechanisms. Most evidence suggests that proximal tubule calcium transport is dominated by passive diffusion with a small component of solvent drag mechanisms. Together, these two mechanisms account for some 90% of proximal tubule calcium absorption. Furthermore, these passive transport mechanisms would dictate a paracellular route for absorption.

Recent information suggests that calcium uptake in proximal tubules may also be mediated by the multifunctional endocytic receptor, megalin [1]. At present, it is unclear whether calcium entering the cell in this fashion is destined for basolateral extrusion and net absorption or subserves some other cytosolic process such as replenishing subcellular calcium stores.

The distinguishing feature of calcium absorption by thick ascending limbs is the capacity for hybrid, parallel transcellular and paracellular transport (fig. 2). Sodium absorption by thick limbs is accomplished by apical membrane Na-K-2Cl cotransport. The characteristic lumen-positive voltage in thick limbs arises from the secretion and recycling of K^+ through apical membrane ROMK channels [2]. Calcium permeability is high and the lumen-positive voltage drives calcium absorption through the paracellular pathway. Increases in sodium absorption are attended by parallel increases in calcium absorption and decreases in sodium transport are accompanied by diminished calcium transport. This is important because administration of 'loop' diuretics that block Na-K-2Cl cotransport reduce the transepithelial voltage and cause parallel reductions in calcium absorption.

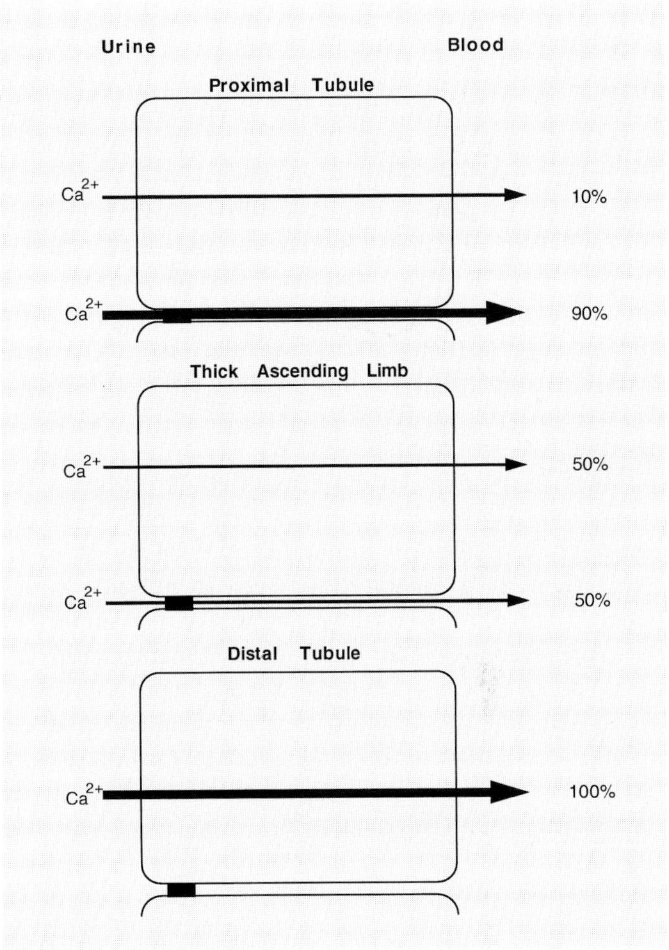

Fig. 2. Relative magnitude of cellular and paracellular calcium absorption in proximal tubules, thick ascending limbs, and distal tubules. In thick limbs, transcellular calcium absorption under maximally stimulated conditions approximates the extent of paracellular calcium transport.

A recurring theme in contemporary biomedical research is the extent to which physiological principles are illuminated by inherited diseases. A case in point is Bartter's syndrome, which is characterized by hypokalemic metabolic alkalosis with renal salt wasting and elevated calcium excretion. Mutations of the gene encoding the Na-K-2Cl cotransporter, *SLC12A1,* are responsible for one form of Bartter's syndrome [3]; others are caused by inactivating mutations of the *CLCNKB* chloride channel gene and *KCNJ1*, which encodes the ATP-sensitive ROMK K^+ channel [3, 4]. Patients with Bartter's syndrome exhibit hypokalemic metabolic alkalosis with renal salt wasting and generally exhibit elevated calcium excretion regardless of which gene is mutated. Thus, sodium and calcium excretion change in parallel, consistent with a derangement in sodium transport by thick ascending limbs. However, a striking difference occurs between patients with ClC-Kb mutations compared with those harboring Na-K-2Cl cotransporter or ROMK mutations. Nephrocalcinosis is absent in patients with ClC-Kb mutations while it is almost universal in patients with Na-K-2Cl or ROMK muations. The precise basis for the different phenotype is unclear and the deviation from the general rule of sodium and calcium excretion being mutually affected suggests that a downstream site may exert a compensatory calcium-absorbing function when the Na-K-2Cl cotransporter or ROMK are involved but not in the case of ClC-Kb mutations.

Thick limbs also possess a transcellular pathway for calcium absorption. This appears to be quiescent under resting conditions but is stimulated by PTH and by calcitonin, though the magnitude of such stimulation varies between species [5]. Little is known of the molecular structures responsible for hormone-mediated calcium entry or about basolateral efflux. Several reports describe poorly selective or nonselective cation channels in apical membranes of thick limbs [6, 7]. Cellular calcium egress is likely to be mediated by one or more of the PMCA isoforms expressed in thick limbs [8]. Although porcine cortical ascending limb cells express the *NCX*1 Na^+/Ca^{2+} exchanger and exchange transport activity [9, 10], *NCX*1 immunofluorescence and single nephron RT-PCR analyses failed to uncover exchanger protein in cortical thick limbs of several species [11–14].

The magnitude of calcium absorption through the paracellular pathway compared to that transported through the cell is uncertain. Some of this uncertainty is due to the complex nature of hormonal stimulation in medullary versus cortical thick limbs, which varies between species. In humans, cortical and medullary thick limbs are equivalently sensitive to PTH but the former are somewhat more responsive to calcitonin than are medullary thick limbs [15]. Thus, the balance of passive and active transport will be determined by the hormonal status and the spontaneous rate of salt absorption along the entire length of the thick ascending limb. The finding that administration of loop diuretics profoundly increases calcium excretion, however, makes it is likely that paracellular calcium transport is the major route of absorption.

Considerably more detail is known of the cellular mechanism of calcium transport by distal tubules, where 8–10% of the filtered calcium is reabsorbed under the hormonal control of PTH, $1,25(OH)_2D_3$, and calcitonin. Two features distinguish calcium transport by distal tubules: first, absorption is active and proceeds entirely by a transcellular route, and second, calcium and sodium absorption are inversely related. Physiological (e.g., PTH), pharmacological (thiazide diuretics, amiloride), pathophysiological (phosphate depletion, adrenocortical insufficiency, metabolic acidosis), and genetic (Gitelman's syndrome) conditions, however, subvert the normal inverse relation. Every known instance of such a dissociation of renal calcium and sodium excretion is attributable to a selective effect on the distal tubule.

As noted above, cellular calcium absorption is a two-step process wherein entry across apical membranes is followed by exit across basolateral cell membranes. Several features of distal calcium entry have been delineated and include the following: in the absence of PTH or other stimulating challenge calcium entry is negligible; calcium entry is blocked by dihydropyridine-type calcium channel blockers [16, 17]; the putative calcium entry channel mediating uptake has a remarkably small single-channel conductance and is activated at hyperpolarizing membrane voltages [17, 18]; the putative channel is multimeric since antisense oligodeoxynucleotides targeted to the calcium channel β subunit expressed by distal tubule cells blocked PTH- and thiazide-induced calcium entry [19]. These findings and others are generally compatible with the view that a single calcium entry mechanism consisting of a multimeric protein complex is responsible for hormone and drug-stimulated calcium entry. However, kinetic analyses [20] and targeted antisense experiments [19] suggest that two distinct mechanisms are involved. According to Lajeunesse et al. [20], a high-affinity, low-capacity system mediates PTH-stimulated calcium entry [20], whereas a low-affinity, high-velocity mechanism mediates thiazide [20] and calcitonin-induced [21] calcium uptake.

Although the functional features help define the behavior of the calcium entry channel, its molecular structure remained elusive until the recent studies by Hoenderop et al. [22, 23]. These investigators used an expression cloning strategy to isolate a cDNA that has an open reading frame of 2,190 nucleotides with a predicted mass of 83 kD that encodes a calcium channel. The channel, an epithelial calcium channel (EcaC), has six putative transmembrane-spanning domains and consensus sites for potential phosphorylation by protein kinase A and C. Both amino and carboxy termini are predicted to be intracellular. ECaC bears no resemblance to other epithelial ion channels though it shares some similarity to the capsaicin receptor, which is a calcium-permeable, nonselective cation channel [24]. One such capsaicin receptor, with which ECaC shares 50% overall sequence identity, was recently cloned from rat kidney [25]. When expressed in *Xenopus laevis* oocytes, the ECaC was spontaneously active and exhibited calcium entry with a K_m of 0.2 m*M*, consistent with that in intact rabbit kidney cells and with luminal fluid calcium concentrations. The channel, which was also localized to the small intestine and placenta, was blocked by La > Cd > Mn > Ba but not by Mg or Sr. Subsequent characterization revealed that ECaC is activated at hyperpolarizing voltages, consistent with the earlier patch clamp studies, exhibited calcium-dependent regulation, and desensitization during repetitive stimulation [23]. Curiously, the channel was not inhibited by either verapamil or felodipine. Single-channel properties have not yet been defined. Thus, it is somewhat uncertain if the enhanced current observed at hyperpolarizing holding voltages is simply the expected consequence of the increased driving force or actual activation, which would be manifested as greater channel open probability (NPo), as described for distal tubule calcium entry channels [18]. Likewise, it is difficult with such incomplete information to reconcile the multimeric nature of the distal tubule calcium entry channel [19] and its dihydropyridine sensitivity with ECaC. Perhaps additional subunits are necessary to confer the pharmacological characteristics observed in DCT cells.

A calcium transport protein (CaT1) that is structurally related to ECaC was recently reported by Peng et al. [26]. The amino acid sequence of CaT1 is 75% identical to ECaC. CaT1 transcripts were found throughout the intestine, in brain, thymus, and adrenal gland but absent from kidney, heart, liver, lung, spleen, and skeletal muscle.

DCTs exhibit the highest Ca^{2+}-ATPase activity along the nephron [27] and display the strongest immunocytochemical reactivity for PMCA protein expression [12, 28, 29]. The *NCX*1 Na^+/Ca^{2+} exchanger is also present [11, 14, 30, 31]. The precise role of these two calcium exit proteins in mediating basal and stimulated levels of calcium efflux is uncertain.

A model of calcium transport by distal tubule cells that incorporates recent observations is portrayed in figure 3. The features of this model include on apical plasma membranes: electroneutral Na^+ and Cl^- entry mediated by the thiazide-sensitive Na-Cl cotransporter; an amiloride-sensitive Na^+ channel, ENaC, and the recently described calcium channel, ECaC. On basolateral membranes, the Na^+-K^+-ATPase provides the energy for diffusional entry of Na^+ and for secondary active transporters such as the Na^+/Ca^{2+} exchanger, K^+ and Cl^- channels, and the

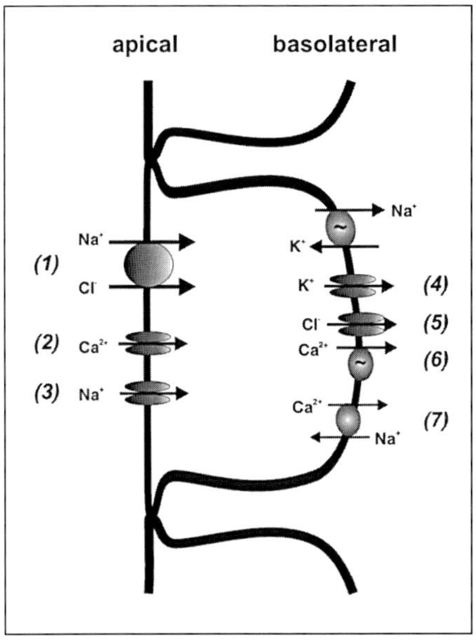

Fig. 3. Working model of distal tubules illustrating the major transporters and channels involved in calcium absorption. Calcium enters the cell across apical membranes through a channel (2) that is activated upon membrane hyperpolarization and exits across basolateral membranes by the Ca^{2+}-ATPase (4) or by Na^+/Ca^{2+} exchange (7). Additional details are given in the text and in table 1.

PMCA. Table 1 lists the different transporters, their genes, gene products, and human chromosome locations.

What then is the mechanism responsible for the inverse relationship between sodium and calcium that is the hallmark of their transport by distal tubules? The primary ion transport proteins in distal tubules (fig. 3) include two apical membrane sodium entry pathways: the Na-Cl cotransporter and the amiloride sensitive epithelial Na^+ channel. Inhibition of either of these hyperpolarizes the membrane voltage and activates the apical membrane calcium channel. Basolateral calcium efflux is mediated by Na^+/Ca^{2+} exchange and by a PMCA. The inverse relations between sodium and calcium absorption by distal tubules can be explained by this model. The common denominator of this scheme is that membrane hyperpolarization activates the apical membrane calcium entry channel [18, 23]. Diminished sodium entry through either the Na-Cl cotransporter [32, 33] or the Na^+ channel [32, 34, 35] hyperpolarizes the membrane. Likewise, PTH [18, 36] and calcitonin acting through their cognate receptors [37] hyperpolarize DCT membrane voltage. Hyperpolarization of membrane voltage by PTH is a direct effect since this action is present even when calcium entry is blocked by nifedipine [36]. Thus, according to this model, sodi-

Table 1. Distal tubule transport proteins and genes

	Transporter	Name	Gene	Locus	Reference
Apical membrane					
1	Na-Cl cotransporter	TSC1	*SLC12A3*	16q13	44
2	Na channel	ENaC	*SCNN1A* *SCNN1B* *SCNN1G*	12p13	45
3	Ca channel	ECaC			22
Basolateral membrane					
4	K^+ channel	ROMK	*KCNJ1*	11q24	46
5	Cl^- channel	ClC-K1 ClC-K2	*CLCNKA* *CLCNKB*	1p36	47
6	Ca^{2+}-ATPase	PMCA1	*ATP2B1*	12q21-q23	48
7	Na/Ca exchanger	NCX1	*SLC8A1*	2p23-p22	49

Numbers 1–7 refer to the transporters depicted in figure 3. Entries are not inclusive but representative of major gene products expressed in distal tubules. Loci are to the human chromosome.

um-dependent changes in calcium transport are indirect and occur secondarily through effects on membrane voltage. Membrane hyperpolarization would also enhance the rate of calcium efflux that is mediated by Na^+/Ca^{2+} exchange, where 3 Na^+ ions enter the cell in exchange of 1 Ca^{2+} [38].

Gitelman's Syndrome and Calcium Transport by Distal Tubules

Just as Bartter's syndrome illustrates the parallel nature of sodium and calcium movement in thick limbs, a variant of Bartter's syndrome described by Gitelman et al. [39] underscores the deductions regarding the inverse nature of sodium and calcium movement in distal tubules. Patients affected with Gitelman's syndrome exhibit salt wasting, hypokalemia, and hypermagnesuria as in Bartter's syndrome but, in contrast, display a conspicuous decrease in calcium excretion [39, 40]. Gitelman's syndrome is due to inactivating mutations of the apical membrane Na-Cl cotransporter gene, *Slc12a3*, which is located on human chromosome 16 [41] (table 1). Not surprisingly, Gitelman patients are refractory to thiazide diuretics since the Na-Cl cotransporter is constitutively inhibited but they remain sensitive to loop diuretics [42]. A transgenic mouse model of Gitelman's syndrome, where the Na-Cl cotransporter was inactivated by homologous recombination, similarly displays reduced calcium excretion [43]. The hypocalciuria of Gitelman's syndrome can be understood from the model of distal tubular calcium absorption depicted in figure 3 and presumably arises from membrane hyperpolarization and persistent stimulation of calcium absorption. Thus, impaired or diminished Na-Cl cotransport enhances calcium absorption. Thus, the disordered mineral ion metabolism accompanying Gitelman's syndrome reinforces the conclusions drawn regarding the cellular mechanisms of distal calcium absorption and the inverse relations between calcium and sodium transport by distal tubules.

Conclusion

Although our understanding of the mechanisms and regulation of transcellular calcium absorption in the kidney remains incomplete, recent findings augment the emerging model for transcellular calcium transport. As the complete array of calcium-transporting and calcium-regulating proteins is cloned, it will become increasingly important to evaluate how they interact in vivo by performing classical physiological experiments including clearance and isolated tubule studies in transgenic animals.

Acknowledgements

Studies from the author's laboratory were supported by NIH grant DK54171. The author is grateful to his many collaborators, especially Drs. Frank Gesek, Libby Barry, Ken White, and Bruce Stanton, and to Bonnie Coutermarsh and Susan Kennedy for their scientific and technical contributions.

References

1 Christensen EI, Birn H, Verroust P, Moestrup SK: Megalin-mediated endocytosis in renal proximal tubule. Ren Fail 1998;20:191–199.

2 Greger R, Bleich M, Schlatter E: Ion channel regulation in the thick ascending limb of the loop of Henle. Kidney Int 1991;40(suppl 33): 119–124.

3 Simon DB, Bindra RS, Mansfield TA, Nelson-Williams C, Mendonca E, Stone R, Schurman S, Nayir A, Alpay H, Bakkaloglu A, Rodriguez-Soriano J, Morales JM, Sanjad SA, Taylor CM, Pilz D, Brem A, Trachtman H, Griswold W, Richard GA, John E, Lifton RP: Mutations in the chloride channel gene, *CLCNKB*, cause Bartter's syndrome type III. Nat Genet 1997; 17:171–178.

4 Simon DB, Karet FE, Rodriguez-Soriano J, Hamdan JH, DiPietro A, Trachtman H, Sanjad SA, Lifton RP: Genetic heterogeneity of Bartter's syndrome revealed by mutations in the K^+ channel, ROMK. Nat Genet 1996;14: 152–156.

5 Friedman PA, Gesek FA: Cellular calcium transport in renal epithelia: Measurement, mechanisms, and regulation. Physiol Rev 1995;75:429–471.

6 Lau K, Quamme G, Tan S: Patch-clamp evidence for a Ca channel in apical membrane of cortical thick ascending limb (cTAL) and distal tubule (DT) cells (abstract). J Am Soc Nephrol 1991;2:775.

7 Chraïbi A, Van den Abbeele T, Guinamard R, Teulon J: A ubiquitous non-selective cation channel in the mouse renal tubule with variable sensitivity to calcium. Pflügers Arch 1994;429: 90–97.

8 Caride AJ, Chini EN, Homma S, Penniston JT, Dousa TP: mRNA encoding four isoforms of the plasma membrane calcium pump and their variants in rat kidney and nephron segments. J Lab Clin Med 1998;132:149–156.

9 Dai LJ, Ritchie G, Bapty B, Raymond L, Quamme GA: Na^+/Ca^{2+} exchanger in epithelial cells of the porcine cortical thick ascending limb. Am J Physiol 1996;270:F411–F418.

10 Dai LJ, Ritchie G, Bapty B, Auger V, Quamme GA: Modulation of Na^+/Ca^{2+} exchange in epithelial cells of porcine thick ascending limb. Am J Physiol 1996;270:F953–F959.

11 Reilly RF, Shugrue CA, Lattanzi D, Biemesderfer D: Immunolocalization of the Na^+/Ca^{2+} exchanger in rabbit kidney. Am J Physiol 1993; 265:F327–F332.

12 Lytton J, Lee SL, Lee WS, van Baal J, Bindels RJM, Kilav R, Naveh-Many T, Silver J: The kidney sodium-calcium exchanger. Ann NY Acad Sci 1996;779:58–72.

13 Bourdeau JE, Taylor AN, Iacopino AM: Immunocytochemical localization of sodium-calcium exchanger in canine nephron. J Am Soc Nephrol 1993;4:105–110.

14 Yu ASL, Hebert SC, Lee S-L, Brenner BM, Lytton J: Identification and localization of renal Na^+-Ca^{2+} exchanger by polymerase chain reaction. Am J Physiol 1992;263:F680–F685.

15 Chabardès D, Gagnan-Brunette M, Imbert-Teboul M, Gontcharevskaia O, Montégut M, Clique A, Morel F: Adenylate cyclase responsiveness to hormones in various portions of the human nephron. J Clin Invest 1980;65:439–448.

16 Bacskai BJ, Friedman PA: Activation of latent Ca^{2+} channels in renal epithelial cells by parathyroid hormone. Nature 1990;347:388–391.

17 Tan S, Lau K: Patch-clamp evidence for calcium channels in apical membranes of rabbit kidney connecting tubules. J Clin Invest 1993; 92:2731–2736.

18 Matsunaga H, Stanton BA, Gesek FA, Friedman PA: Epithelial Ca^{2+} channels sensitive to dihydropyridines and activated by hyperpolarizing voltages. Am J Physiol 1994;267:C157–C165.

19 Barry ELR, Gesek FA, Yu ASL, Lytton J, Friedman PA: Distinct calcium channel isoforms mediate parathyroid hormone and chlorothiazide-stimulated calcium entry in transporting epithelial cells. J Membr Biol 1998; 161:55–64.

20 Lajeunesse D, Bouhtiauy I, Brunette MG: Parathyroid hormone and hydrochlorothiazide increase calcium transport by the luminal membrane of rabbit distal nephron segments through different pathways. Endocrinology 1994;134:35–41.

21 Zuo Q, Claveau D, Hilal G, Leclerc M, Brunette MG: Effect of calcitonin on calcium transport by the luminal and basolateral membranes of the rabbit nephron. Kidney Int 1997;51: 1991–1999.

22 Hoenderop JGJ, van der Kemp AWCM, Hartog A, van de Graaf SFJ, van Os CH, Willems PHGM, Bindels RJ: Molecular identification of the apical Ca^{2+} channel in 1,25-dihydroxyvitamin D_3-responsive epithelia. J Biol Chem 1999;274:8375–8378.

23 Hoenderop JGJ, van der Kemp AWCM, Hartog A, van Os CH, Willems PHGM, Bindels RJ: The epithelial calcium channel, ECaC, is activated by hyperpolarization and regulated by cytosolic calcium. Biochem Biophys Res Commun 1999;261:488–492.

24 Caterina MJ, Schumacher MA, Tominaga M, Rosen TA, Levine JD, Julius D: The capsaicin receptor: A heat-activated ion channel in the pain pathway. Nature 1997;389:816–824.

25 Suzuki M, Sato J, Kutsuwada K, Ooki G, Imai M: Cloning of a stretch-inhibitable nonselective cation channel. J Biol Chem 1999;274: 6330–6335.

26 Peng JB, Chen XZ, Berger UV, Vassilev PM, Tsukaguchi H, Brown EM, Hediger MA: Molecular cloning and characterization of a channel-like transporter mediating intestinal calcium absorption. J Biol Chem 1999;274: 22739–22746.

27 Doucet A, Katz AI: High-affinity Ca-Mg-ATPase along the rabbit nephron. Am J Physiol 1982;242:F346–F352.

28 Borke JL, Minami J, Verma A, Penniston JT, Kumar R: Monoclonal antibodies to human erythrocyte membrane Ca^{++}-Mg^{++} adenosine triphosphatase pump recognize an epitope in the basolateral membrane of human kidney distal tubule cells. J Clin Invest 1987;80:1225–1231.

29 Borke JL, Caride A, Verma AK, Penniston JT, Kumar R: Plasma membrane calcium pump and 28-kDa calcium binding protein in cells of rat kidney distal tubules. Am J Physiol 1989; 257:F842–F849.

30 Ramachandran C, Brunette MG: The renal Na^+/Ca^{2+} exchange system is located exclusively in the distal tubule. Biochem J 1989;257: 259–264.

31 White KE, Gesek FA, Reilly RF, Friedman PA: *NCX*1 Na/Ca exchanger inhibition by antisense oligonucleotides in mouse distal convoluted tubule cells. Kidney Int 1998;54:897–906.

32 Stanton BA: Cellular actions of thiazide diuretics in the distal tubule. J Am Soc Nephrol 1990; 1:832–836.

33 Gesek FA, Friedman PA: Mechanism of calcium transport stimulated by chlorothiazide in mouse distal convoluted tubule cells. J Clin Invest 1992;90:429–438.

34 Friedman PA, Gesek FA: Stimulation of calcium transport by amiloride in mouse distal convoluted tubule cells. Kidney Int 1995;48: 1427–1434.

35 Dai L-J, Friedman PA, Quamme GA: Mechanisms of amiloride stimulation of Mg^{2+} uptake in immortalized mouse distal convoluted tubule cells. Am J Physiol 1997;272:F249–F256.

36 Gesek FA, Friedman PA: On the mechanism of parathyroid hormone stimulation of calcium uptake by mouse distal convoluted tubule cells. J Clin Invest 1992;90:749–758.

37 Gesek FA, Friedman PA: Calcitonin stimulates calcium transport in distal convoluted tubule cells. Am J Physiol 1993;264:F744–F751.

38 Friedman PA: Codependence of renal calcium and sodium transport. Annu Rev Physiol 1998; 60:179–197.

39 Gitelman HJ, Graham JB, Welt LG: A new familial disorder characterized by hypokalemia and hypomagnesemia. Trans Assoc Am Physicians 1966;79:221–235.

40 Bettinelli A, Bianchetti MG, Girardin E, Caringella A, Cecconi M, Appiani AC, Pavanello L, Gastaldi R, Isimbaldi C, Lama G, et al : Use of calcium excretion values to distinguish two forms of primary renal tubular hypokalemic alkalosis: Bartter and Gitelman syndromes. J Pediatr 1992;120:38–43.

41 Simon DB, Nelson-Williams C, Bia MJ, Ellison D, Karet FE, Molina AM, Vaara I, Iwata F, Cushner HM, Koolen M, Gainza FJ, Gitelman HJ, Lifton RP: Gitelman's variant of Bartter's syndrome, inherited hypokalaemic alkalosis, is caused by mutations in the thiazide-sensitive Na-Cl cotransporter. Nat Genet 1996;12:24–30.

42 Colussi G, Rombolà G, Brunati C, De Ferrari ME: Abnormal reabsorption of Na^+/Cl^- by the thiazide-inhibitable transporter of the distal convoluted tubule in Gitelman's syndrome. Am J Nephrol 1997;17:103–111.
43 Schultheis PJ, Lorenz JN, Meneton P, Nieman ML, Riddle TM, Flagella M, Duffy JJ, Doetschman T, Miller ML, Shull GE: Phenotype resembling Gitelman's syndrome in mice lacking the apical Na^+-Cl^- cotransporter of the distal convoluted tubule. J Biol Chem 1998; 273:29150–29155.
44 Gamba G, Saltzberg SN, Lombardi M, Miyanoshita A, Lytton J, Hediger MA, Brenner BM, Hebert SC: Primary structure and functional expression of a cDNA encoding the thiazide-sensitive, electroneutral sodium-chloride cotransporter. Proc Natl Acad Sci USA 1993;90: 2749–2753.
45 Canessa CM, Horisberger JD, Rossier BC: Epithelial sodium channel related to proteins involved in neurodegeneration. Nature 1993; 361:467–470.
46 Ho K, Nichols CG, Lederer WJ, Lytton J, Vassilev PM, Kanazirska MV, Hebert SC: Cloning and expression of an inwardly rectifying ATP-regulated potassium channel. Nature 1993; 362:31–38.
47 Kieferle S, Fong P, Bens M, Vandewalle A, Jentsch TJ: Two highly homologous members of the ClC chloride channel family in both rat and human kidney. Proc Natl Acad Sci USA 1994;91:6943–6947.
48 Nicoll DA, Longoni S, Philipson KD: Molecular cloning and functional expression of the cardiac sarcolemmal Na^+-Ca^{2+} exchanger. Science 1990;250:562–565.
49 Reilly RF, Shugrue CA: cDNA cloning of a renal Na^+-Ca^{2+} exchanger. Am J Physiol 1992; 262:F1105–F1109.

Experimental Nephrology

Exp Nephrol 2000;8:351–360

Molecular Pathology of Renal Chloride Channels in Dent's Disease and Bartter's Syndrome

R.V. Thakker

Molecular Endocrinology Group, University of Oxford, John Radcliffe Hospital, Headington, Oxford, Oxon, UK

Key Words
Chloride channel gene · Hypercalciuria · Low molecular weight proteinuria · Kidney stones

Abstract
Recent advances in molecular biology have characterised a new class of chloride channels that are referred to as voltage-gated chloride channels (CLCs). To date 9 such CLCs (CLC-1 to CLC-7, CLC-Ka and CLC-Kb which are respectively encoded by the genes CLCN1 to CLCN7, CLCNKa and CLCNKb) have been identified in mammals. Mutations in 2 of these, referred to as CLC-5 and CLC-Kb, have been defined in the hypercalciuric nephrolithiasis disorders of Dent's disease and a form of Bartter's syndrome, respectively. In addition, other forms of Bartter's syndrome have been defined with mutations involving the bumetanide-sensitive sodium-potassium-chloride co-transporter (NKCC2) and the potassium channel ROMK. Finally, mutations of the thiazide-sensitive sodium chloride co-transporter (NCCT) are associated with Gitelman's syndrome, in which hypocalciuria and hypomagnesaemia are notable features. These molecular genetic studies have increased our understanding of the renal tubular mechanisms that regulate mineral homeostasis.

Recent studies have shed light on some of the renal tubular mechanisms regulating electrolyte homeostasis [1–4] and, in particular, those that regulate chloride which is the most abundant anion in the extracellular fluid (ECF) compartment. The renal tubules reclaim more than 99% of the 180 litres of fluid filtered through the glomeruli each day, and a directional transport of ions and water across the renal tubular cells from the apical (lumen) to the basolateral (blood) side (fig. 1) is thus required. The directional transport of chloride involves chloride channels and chloride co-transporters, and our understanding of these has been increased by studies of hereditary disorders of electrolyte homeostasis [3–12]. Thus, chloride transport abnormalities in Dent's disease, Bartter's syndrome and Gitelman's syndrome have been defined, and these will be reviewed.

Dent's Disease

Four disorders of hereditary hypercalciuric nephrolithiasis (kidney stones) that have been referred to as Dent's disease [13, 14], X-linked recessive nephrolithiasis (XRN) [15, 16], X-linked recessive hypophosphataemic rickets (XLRH) [17] and the idiopathic low molecular weight proteinuria of Japanese children (JILMWP) [18]

KARGER
Fax + 41 61 306 12 34
E-Mail karger@karger.ch
www.karger.com

1018–7782/00/0086–0351$17.50/0

Accessible online at:
www.karger.com/journals/exn

Prof. R.V. Thakker
Molecular Endocrinology Group, Level 7, University of Oxford
John Radcliffe Hospital, Headington, Oxford, Oxon OX3 9DU (UK)
Tel. +44 1865 222043, Fax +44 1865 222049
E-Mail rajesh.thakker@ndm.ox.ac.uk

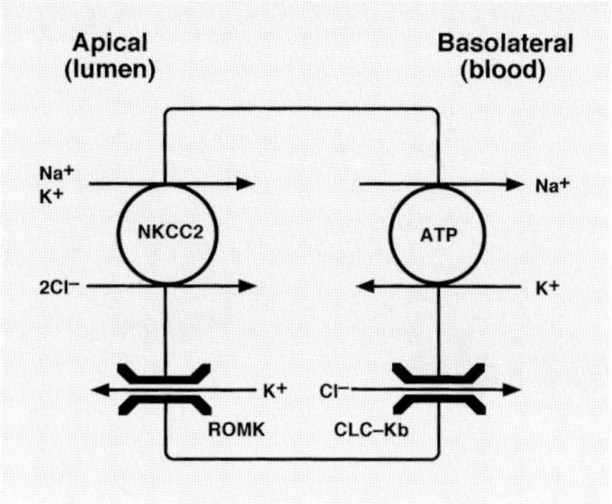

Fig. 1. Schematic representation of a cell from the renal thick ascending limb (TAL) of the loop of Henle. The bumetanide-sensitive Na^+-K^+-Cl^- co-transporter (NKCC2) [7], which consists of 12 transmembrane domains and has significant sequence similarity to the thiazide-sensitive Na^+-Cl^- co-transporter (NCCT) [6], is situated in the apical (luminal side) membrane, and facilitates the entry of Na^+, K^+ and Cl^-. The availability of K^+ in the lumen is rate-limiting for NKCC2 activity, and K^+ entering the TAL cell is recycled into the lumen by the potassium channel ROMK [8, 9] which consists of two putative transmembrane domains, an amphipathic region, a putative pore and an ATP-binding regulatory domain. Exit of Cl^- across the basolateral membrane (blood side) may occur via the chloride channel, CLC-Kb, and exit of Na^+ is via the Na^+-K^+ ATPase (ATP). Mutations of NKCC2, ROMK and CLC-Kb are associated with Bartter's syndrome [3, 4, 7–9].

have been reported to be due to mutations of the X-linked renal specific voltage-gated chloride channel CLC-5 (table 1) [5,10–12, 14, 19, 20]. All these 4 diseases have features in common, and they represent renal tubular disorders that are characterised by low molecular weight proteinuria (LMWP), hypercalciuria, nephrocalcinosis, nephrolithiasis and renal failure. In addition, other renal proximal tubular defects, which include aminoaciduria, phosphaturia, glycosuria, kaliuresis, uricosuria and an acquired impairment of urinary acidification, may also occur [13, 15, 17, 18]. However, there are differences between these disorders; for example, rickets has been a particular feature of Dent's disease and XLRH, but not XRN or JILMWP, and severe renal failure has been a feature of Dent's disease and XRN [13–18]. Dent's disease, XRN and XLRH were mapped to Xp11.22 [14, 16, 17], and a microdeletion in 1 patient with Dent's disease facilitated the isolation and characterisation of a renal chloride channel gene, CLCN5 (fig. 2) [14, 21, 22]. DNA sequence analyses have detected 61 CLCN5 mutations (fig. 2, 3; table 1) that consist of nonsense, missense, splice site, insertional and deletional mutations in patients with these four hypercalciuric nephrolithiasis disorders [5, 10–12, 19, 20], together with another form of a hereditary renal tubular acidosis [23], thereby establishing its casual role in these diseases [24, 25]. In addition, heterologous expression of the wild-type CLCN5 gene or its mutants in *Xenopus* oocytes demonstrated that the wild-type channel, CLC-5, conducted outwardly directed chloride (Cl^-) currents that were either abolished or markedly reduced by the mutations, thereby demonstrating the functional importance of such mutations [5, 10, 11]. The 61 reported mutations (fig. 2, 3; table 1) are scattered throughout the channel with no evidence for mutational hotspots. Furthermore, there appears to be no correlation between the mutations and phenotypes [5, 25]. Of the total 61 CLCN5 mutations approximately 31% are nonsense mutations, 26% are frameshift deletions or insertions, 2% are in-frame insertions, 3% are donor splice site mutations, 5% are acceptor splice site mutations, 26% are missense mutations, 5% are intragenic deletions and 2% are complete deletions of the gene. The majority (>70%) are predicted to result in truncated or absent CLC-5s (table 1) which would lead to a complete loss of channel function. Only 4 of the mutations (30: ins His, G57V, L278F and R280P), which are predicted not to result in truncated CLC-5s and which are outside the predicted transmembrane domains (fig. 3), have been reported to be associated with reduced, but not abolished Cl^- currents. The common genetic aetiology of CLCN5 mutations and the phenotypic similarities between all these syndromes indicated that they were variants of one disorder, and it has been proposed to refer to them collectively as Dent's disease [11, 12].

CLC Gene Family

The CLCs represent the most recently discovered of the three structurally defined classes of chloride channels [4]; these are the extracellular ligand-gated (ELG) receptor chloride channels, e.g. the glycine and γ-aminobutyric acid ($GABA_A$) receptors [26, 27], the cystic fibrosis transmembrane conductance regulator (CFTR) Cl^- channel [28] and the CLCs. The glycine and $GABA_A$ receptors, which probably have 4 transmembrane domains and may function as pentamers, conduct chloride ions and are mainly involved in neuronal inhibition. The CFTR, which is a member of the super-family of ATP-binding cassette (ABC) transporters and has 12 putative transmembrane domains arranged in 2 separate blocks with 2 cyclic nucleotide binding domains and a regulatory region, functions as a cAMP-regulated chloride channel. The CLCs, which are structurally unrelated to the other ion channels and form the only known large family (ta-

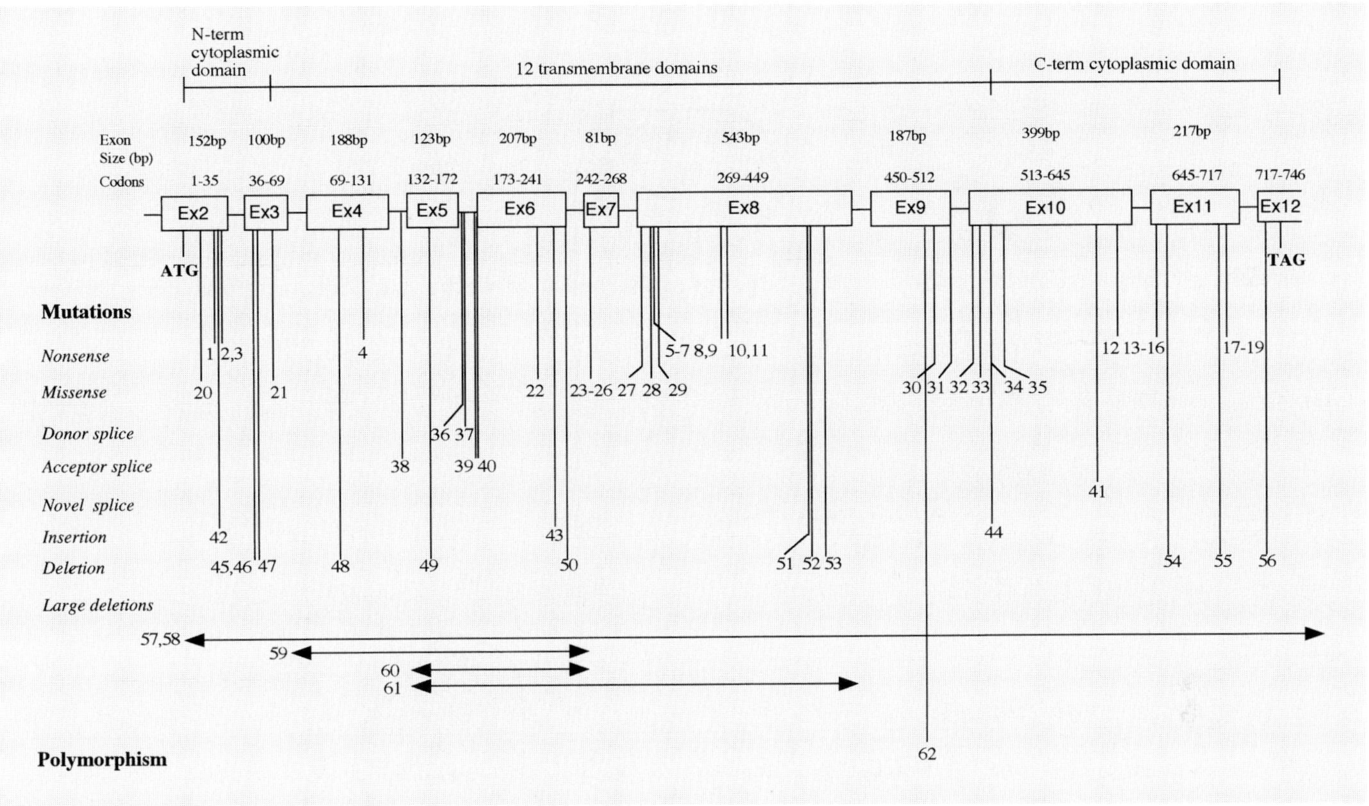

Fig. 2. Schematic representation of the genomic structure of the CLCN5 gene. The human CLCN5 gene consists of 12 exons that span 25–30 kb of genomic DNA [10, 12] and encodes a 746-amino acid protein (fig. 3). The coding region is organised into 11 exons (Ex2 to Ex12) and 10 introns (indicated by the line but not to scale). The gene encodes a chloride channel with 12 transmembrane domains (fig. 3). Exon 2 and part of exon 3 encode the 57 amino acids of the NH_2-terminal cytoplasmic domain; the 3′ end of exon 3 and exon 4 to part of exon 10 encode the 491 amino acids of the transmembrane domains and loops; and the 3′ end of exon 10, exon 11, and part of exon 12 encode the 198 amino acids of the COOH-terminal cytoplasmic domain. The start (ATG) and stop (TAG) sites in exons 2 and 12, respectively, are indicated. The sites of the 61 mutations: 19 nonsense, 16 missense, 2 donor splice site, 3 accepter splice site, 1 novel splice site, 3 insertional, and 17 deletional (12 small and 5 large) mutations and 1 neutral polymorphism are shown below; the details of each of these are provided in table 1.

ble 2) of Cl^- channels, consist of about 12 transmembrane domains (fig. 3) [5, 21, 22, 29, 30]. The correct number and topology of these domains is being established [31, 32], but domain 3 is likely to be a transmembrane domain [32], and recent studies of domain 4 using cysteine-scanning mutagenesis indicate that domain 4, which contains a conserved motif that is essential for anion selectivity, is likely to be involved in pore formation [31]. The first member, designated CLC-0, was cloned in 1990 from the electric organ of *Torpedo marmorata* [29], and nine different CLCs (CLC-1 to CLC-7, and CLC-Ka and CLC-Kb) encoded respectively by genes CLCN1 to CLCN7, and CLCNKa and CLCNKb, have been identified in mammals [21, 30, 33]. Heterologous expression in *Xenopus* oocytes has revealed that the chloride channels CLC-0, CLC-1, CLC-2 and CLC-5 conduct chloride currents that are outwardly rectifying and time-dependent, and with a conductivity sequence that prefers chloride to iodide [5, 10, 29–32, 34, 35]. These chloride channels are important for the control of membrane excitability, transepithelial transport and possibly regulation of cell volume [30]. CLCs are known to function as multimeric complexes and recent studies have revealed that CLC-0 is a homodimer with two largely independent pores [34, 35]. The CLC genes are expressed in a variety of tissues (table 2) and only CLCN1, CLCN5 and CLCNKb have to date been reported to have disease-associated mutations in man [4]. Thus, mutations of CLCN1 are associated with the myotonia disorders of Thomsen and Becker [36–38]; mutations of CLCN5 are associated with the hereditary neph-

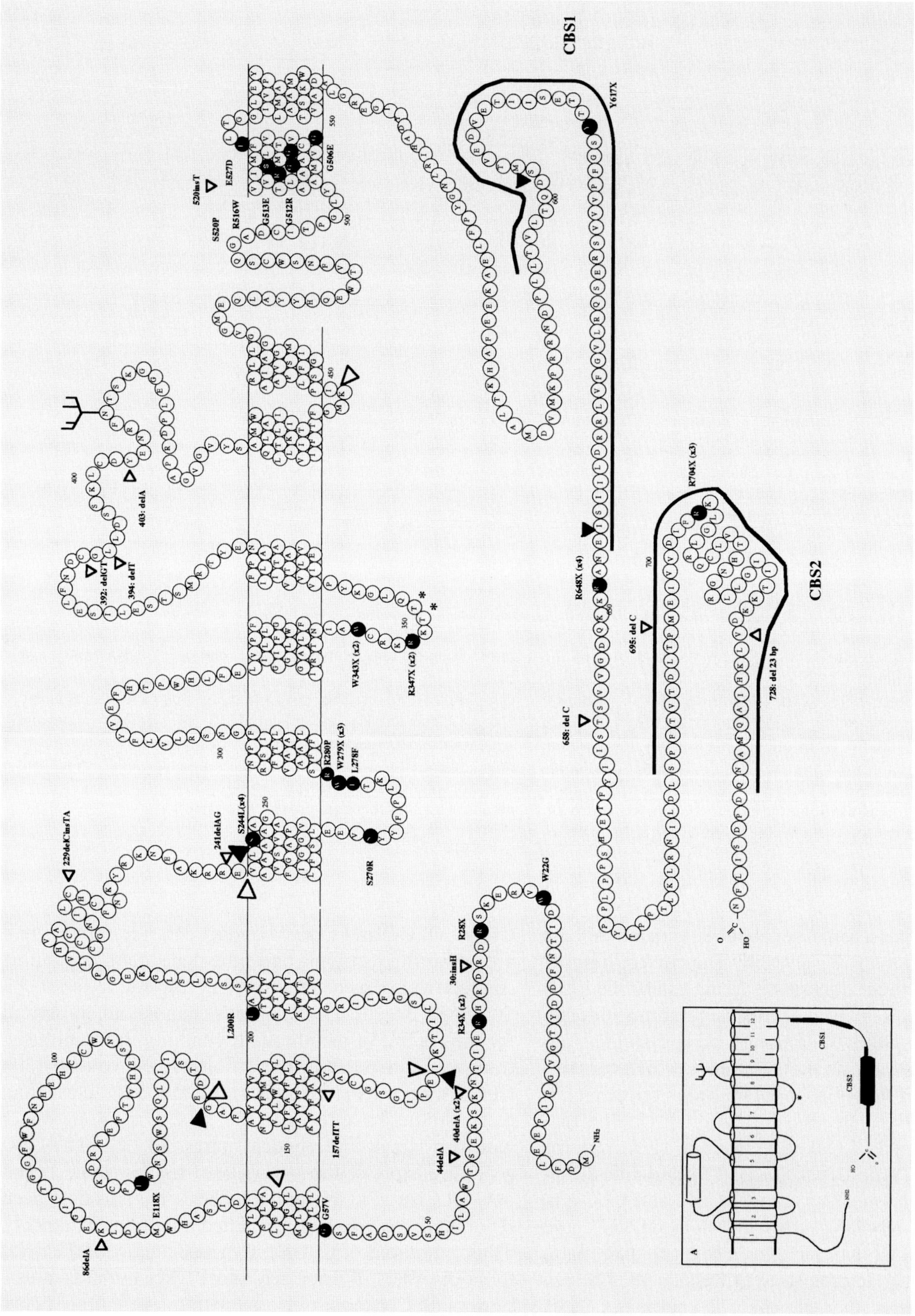

CBS1
CBS2
86delA
E118X
G57V
44delA
40delA (x2)
157delTT
L200R
R34X (x2)
30:insH
R28X
W22G
229delC'insTA
241delAG
S244L(x4)
S270R
R280P
W279X (x3)
L278F
W343X (x2)
R347X (x2)
392: delGT
394: delT
403: delA
520insT
S520P
R516W
E527D
G513E
G512R
G506E
Y617X
R648X (x4)
658: del C
695: del C
728: del 23 bp
R704X (x3)

Table 1. Mutations in the CLCN5 gene

Number[a]	Codon	Mutation[b]	Predicted effect[c]	Reference
Nonsense				
1	28	CGA → TGA	R28X	56
2	34	CGA → TGA	R34X	56
3	34	CGA → TGA	R34X	40
4	118	GAG → TAG	E118X	57
5	279	TGG → TGA	W279X	10
6	279	TGG → TGA	W279X	5
7	279	TGG → TGA	W279X	12
8	343	TGG → TGA	W343X	10
9	343	TGG → TGA	W343X	56
10	347	CGA → TGA	R347X	12
11	347	CGA → TGA	R347X	57
12	617	TCA → TAA	Y617X	41
13	648	CGA → TGA	R648X	5
14	648	CGA → TGA	R648X	11
15	648	CGA → TGA	T648X	58
16	648	CGA → TGA	R648X	40
17	704	CGA → TGA	R704X	5
18	704	CGA → TGA	R704X	40
19	704	CGA → TGA	R704X	58
Missense				
20	22	TGG → GGG	W22G	57
21	57	GGC → GTC	G57V	11
22	200	CTG → CGG	L200R	5
23	244	TCG → TTG	S244L	5
23	244	TCG → TTG	S244l	23
25	244	TCG → TTG	S244L	19
26	244	TCG → TTG	S244L	56
27	270	AGC → CGC	S270R	58
28	278	TTG → TTC	L278F	58
29	280	CGT → CCT	R280P	10
30	506	GGG → GAG	G506E	5
31	512	GGT → CGT	G512R	11
32	513	GGG → GAG	G513E	12
33	516	CGG → TFF	R516W	12
34	520	TCT → CCT	S520P	5
35	527	GAA → GAT	G527D	11
Donor splice site				
36	intron 5	gtgag → atgag	exon 5 skipped	5
37	intron 5	gtgag → gggag	exon 5 skipped	5
Acceptor splice site				
38	intron 4	tgcag → tgccg	?	58
39	intron 5	tgcag → tgcgg	exon 6 skipped, or cryptic splice site used in exon 6, all result in truncated channels	59
40	intron 5	tgcag → tgaag	exon 6 skipped	40
Novel splice site				
41	601	nt 1802 (A → T)	missense peptide from codon 601–632 and loss of remaining 114 amino acids	41

➤

Fig. 3. Schematic representation of a predicted topology of CLC-5. The mutations associated with Dent's disease, and its phenotypic variants are illustrated using the reported model [5] for CLC-5 topology. Every 50th amino acid of the 746-amino acid CLC-5 protein and the consensus phosphorylation (asterisk) and glycosylation (branch) sites at codons 349, 350 and 408, respectively, are indicated. In addition, the C-terminal segment is thought to contain two cystathionine β-synthase (CBS) domains [50–53]; CBS domains in inosine-5′-monophosphate dehydrogenase (IMPDH) and the yeast SNF4 are known to provide binding sites for interacting molecules [54, 55]. The CLC-5 mutations are shown with the amino acid highlighted in black and the codon and substituted amino acid shown alongside. The correct topology for the CLC channels is unknown and the predicted topology of the CLC-5 putative transmembrane domains (D1–D13) is based upon a model [5] (inset) that places D4 extracellularly, and in which the hydrophobic core of the D9–D12 region crosses the membrane 3 or 5 times and contains a hydrophilic region (codons 481–502), the precise location of which remains unknown. More recent studies indicated that D4 is likely to be involved in pore formation [31]. Of the 61 CLCN5 mutations reported (table 1), 45 are shown and 8 of these have been observed to occur more than once (twice to four times); in addition, 3 intragenic deletions (open large triangles) and 3 splice site mutations (filled triangles) are shown. Such CLCN5 mutations have been demonstrated by heterologous expression in *Xenopus* oocytes to result in a functional loss of the channel CLC-5 [5, 10, 11].

Table 1 (continued)

Number[a]	Codon	Mutation[b]	Predicted effect[c]	Reference
Insertion				
42	30	ins ACC	if, ins H	11
43	229	del C ins TA	fs	12
44	520	ins T	FS	20
Deletion				
45	40	del A	fs	56
46	40	del A	fs	40
47	44	del A	fs	56
48	86	del A	fs	40
49	157	del TT	fs	40
50	241	del AG	fs	40
51	392	del GT	fs	41
52	394	del T	fs	41
53	403	del A	fs	20
54	658	del C	fs	41
55	695	del C	fs	10
56	728	del 23bp	fs	41
57	>180 kb del		no protein	12
58	515 kb del		no protein	5
59	>0.6 kb		exons 4, 5 and 6 del	59
60	2 kb del		exons 5 and 6 del	5
61	>7 kb del		exons 5, 6, 7 and 8 del	57
Polymorphism				
62	484	AC<u>C</u> → AC<u>T</u>	T484T neutral polymorphism (i.e. no change in amino acid)	12

[a] Number of mutations as indicated by its location in figure 2.
[b] ins = Insertion; del = Deletion.
[c] Predicted effect: amino acid shown by one letter code; X = premature stop codon; ? = not characterised; if = inframe; fs = frameshift with missense peptide, followed by premature stop codon.

rolithiasis disorders of Dent's disease [5, 10, 11], and mutations of CLCNKb are associated with a form of Bartter's syndrome [3, 4]. Recent studies of a mouse knockout model for CLCNK1, which is the murine homologue for the human CLCNKa, have demonstrated the occurrence of a form of nephrogenic diabetes insipidus; this results from a defect in the renal urinary-concentrating mechanism and reveals an important role for CLCNK1 in the counter-current multiplication system of the inner medulla [39]. The human equivalent of this murine model for nephrogenic diabetes insipidus remains to be identified.

Role of CLC-5 in the Nephron

The positional cloning [14, 21] and mutational analysis [5, 40, 41] of Dent's disease and its phenotypic variants indicated that CLCN5 encodes a chloride channel, CLC-5, whose functional loss results in a generalised proximal tubular defect, i.e. Fanconi syndrome, which is associated with the hypercalciuria and nephrolithiasis of Dent's disease [24]. However, the mechanisms whereby a loss of this renal chloride channel (CLC-5) leads to hypercalciuria and the generalised proximal tubular defects remain to be elucidated. Studies of CLC-5 expression in the human nephron have suggested some possibilities as CLC-5 is expressed at multiple sites which include the proximal tubule, thick ascending limb of Henle, and intercalated cells of the collecting duct [42]. CLC-5 is located intracellularly and in the early endosomes, which contain the vacuolar H^+-ATPase and which form part of the receptor-mediated endocytic pathway (fig. 4) [42]. This suggests a role for CLC-5 as a counter ion transport mechanism allowing the action of the electrogenic H^+-ATPase in the acidification of the endosomes [42, 43]. Indeed, CLC-5 has been shown to have a suitable pH dependence for its activity [44] and endosomal acidification is inhibited by non-specific chloride channel blockers. A possible explanation for the proximal tubular re-absorptive defects seen in Dent's disease may then be as follows: low molecular weight proteinuria may be due to defective endocytosis of these proteins [45–47] which are known to be absorbed through the receptor-mediated endocytic pathway, and the other features, e.g., glycosuria and aminoaciduria may be due to defective endosomal recycling of plasma membrane transporters responsible for the sodium-coupled reabsorption of these solutes [42, 43, 48]. The hypercalciuria, which is not a typical feature of proximal tubular

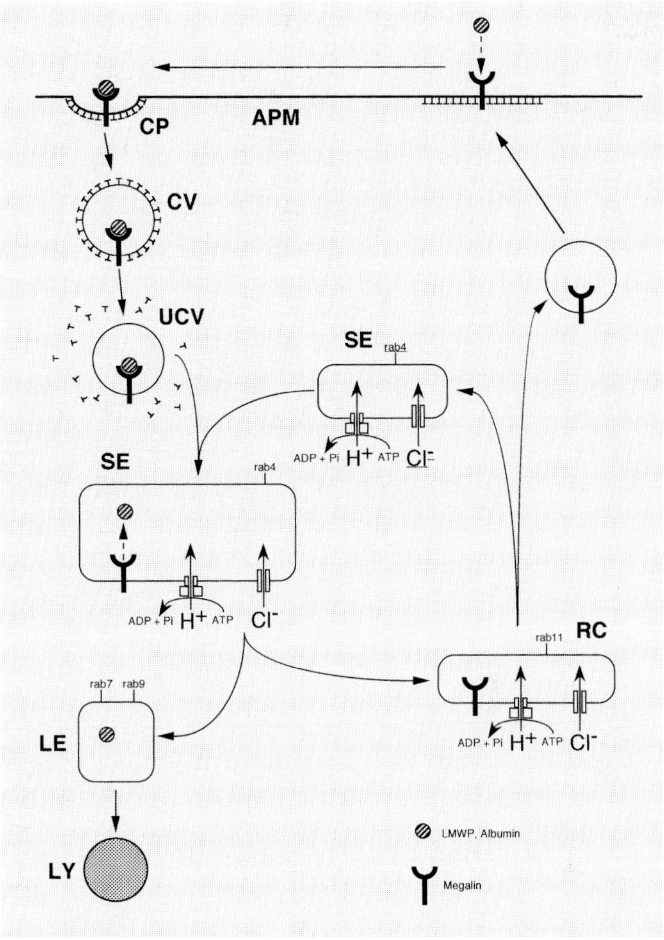

Fig. 4. Schematic representation of the receptor-mediated endocytic pathway in the renal proximal tubular cell [46]. Ligands, e.g. albumin and low molecular weight proteins (LMWP), bind to megalin receptors in clathrin-coated pits (CP) at the apical membrane (APM). The receptor-ligand complex is internalised in a coated vesicle (CV), which is uncoated (UCV) before fusion with a sorting endosome (SE). Acidification of the endosome by the vacuolar H^+-ATPase, with counter ions (Cl^-) probably provided by CLC-5 dissociates ligand from receptor [42, 43]. The receptor is recycled, via the recycling compartment (RC), to the APM, whilst ligand is transported to the late endosomes (LE) and lysosomes (LY), where it is degraded. Rab proteins, which are small membrane-bound GTPases associated with particular endosomal compartments [42] are shown.

Table 2. The mammalian voltage-gated chloride channels (CLCs) and disease associations

Chloride channel	Chromosomal location	Function	Tissue distribution	Disease association
CLC-1	7q35	Voltage stabilisation	Skeletal muscle	Thomsen's myotonia Becker myotonia Arrested development of righting in mouse
CLC-2	3q26–3q28	Cell volume regulation	Ubiquitous	–
CLC-3	4	?	Multiple	–
CLC-4	Xp22.3	?	Muscle, brain, heart	–
CLC-5	Xp11.22	Endosomal chloride transport	Kidney (predominantly)	Dent's disease (nephrolithiasis)
CLC-6	1p36	?	Multiple (e.g. brain, testes, muscle, kidney)	–
CLC-7	16p13	?	Multiple (e.g. brain, testes, muscle, kidney)	–
CLC-Ka	1p36	Concentration gradient in counter-current mechanism	Kidney	Nephrogenic diabetes insipidus in mouse
CLC-Kb	1p36	Chloride reabsorption?	Kidney	Bartter's syndrome

disorders, may partly be of the absorptive type and be due to the associated high normal to elevated serum concentrations of 1,25-dihydroxyvitamin D_3 that are observed in patients with Dent's disease [49]. However, the localisation of CLC-5 to the thick ascending limb of Henle [42], where the majority of calcium re-absorption occurs, suggests another interesting mechanism; the possible changes in endosomal re-cycling stemming from CLC-5 dysfunction in this segment may lead to altered recycling and activities of the calcium-sensing receptor (CaSR) and surface transporters, e.g. the sodium-potassium-chloride co-transporter (NKCCT), the chloride channel (CLC-Kb), and the renal outer medullary potassium channel (ROMK), which are involved in electrolyte (fig. 1) and calcium uptake. The role of Dent's disease-associated CLC-5 mutants (fig. 3) in these physiological mechanisms of calcium homeostasis still remains to be elucidated.

Bartter's and Gitelman's Syndromes

Bartter's syndrome is an autosomal recessive disorder of electrolyte homeostasis characterised by hypokalaemic alkalosis, renal salt wasting that may lead to hypotension, hyper-reninaemic hyperaldosteronism, increased urinary prostaglandin excretion, and hypercalciuria with nephrocalcinosis [3, 7]. However, a subset of such patients with hypokalaemic alkalosis have hypocalciuria with hypomagnesaemia, and this is referred to as Gitelman's syndrome [6]. Kindreds with Gitelman's syndrome may have mutations involving the thiazide-sensitive Na^+-Cl^- (NCCT) co-transporter [6], and studies of kindreds with Bartter's syndrome have demonstrated genetic heterogeneity with mutations involving either the bumetanide-sensitive Na^+-K^+-Cl^- (NKCC2) co-transporter [7] (fig. 1), or the outwardly rectifying renal potassium channel (ROMK) [8, 9], or the voltage-gated chloride channel, CLC-Kb [3] (table 2). Bartter's syndrome patients with NKCC2, ROMK or CLC-Kb mutations have a similar phenotype with one notable exception; nephrocalcinosis was absent in all the 17 patients reported to date with CLC-Kb mutations, despite the occurrence of hypercalciuria in 65% of them [3].

The selective localisation of the chloride transporter, NKCC2, and the chloride channel CLC-Kb on the apical and basolateral membranes, respectively, of the thick ascending limb cells (fig. 1) of Henle's loops enables the directional transport of chloride. Chloride is reabsorbed from the tubular lumen across the apical membrane by the NKCC2 co-transporter, whose activity is maintained by recycling of K^+ by ROMK [8, 9]. This leads to an accumulation of intracellular chloride, which leaves the cell via the chloride channel, CLC-Kb [3] located in the basolateral membrane and enters the blood. Thus, vectorial transport is achieved by the thick ascending limb cell. Interestingly, the opposite arrangement of chloride co-transporters and channels on the basolateral and apical membranes, respectively, would be predicted to lead to chloride secretion rather than re-absorption.

Acknowledgement

I am grateful to the Medical Research Council (MRC) in the United Kingdom for support and to Miss Julie Allen for typing the manuscript, and to Drs P. Christie, B. Harding and A. Nesbitt for help in preparing the figures and tables.

References

1 Scheinmann SJ, Guay-Woodward LM, Thakker RV, Warnock DG: Genetic disorders of renal mineral transport. Engl J Med 1999;340: 1177–1187.
2 Hebert SC, Brown EM, Harris HW: Role of the Ca^{++}-sensing receptor in the divalent mineral ion homeostasis. J Exp Biol 1997;200:295–302.
3 Simon DB, Bindra RS, Mansfield TA, Nelson-Williams C, Mendonca E, Stone R, Shurman S, Nayir A, Alpay H, Aysin B, Rodriguez-Soriano J, Morales JM, Sanjad SM, Taylor CM, Pilz D, Brem A, Trachtman H, Griswold W, Richard GA, John E, Lifton RP: Mutations in the chloride channel gene CLCNKB, cause Bartter's syndrome type III. Nat Genet 1997;17:171–178.
4 Thakker RV: Chloride channels cough up. Nat Genet 1997;17:125–127.
5 Lloyd SE, Pearce SHS, Fisher SE, Steinmeyer K, Schwappach B, Scheinman SJ, Harding B, Bolino A, Devoto M, Goodyer P, Rigden SPA, Wrong OM, Jentsch TJ, Craig IW, Thakker RV: A common molecular basis for three inherited kidney stone diseases. Nature 1996;379: 445–449.
6 Simon DB, Nelson-Williams C, Bia MJ, Ellison D, Karet FE, Molina AM, Vaara I, Iwata F, Cushner HM, Koolen M, Gainza FJ, Gitleman HJ, Lifton RP: Gitelman's variant of Bartter's syndrome, inherited hypokalaemic alkalosis, is caused by mutations in the thiazide-sensitive Na-Cl cotransporter. Nat Genet 1996;12:24–30.
7 Simon DB, Karet FE, Hamdan JM, DiPietro A, Sanjad SA, Lifton RP: Bartter's syndrome, hypokalaemic alkalosis with hypercalciuria, is caused by mutations in the Na-K-2Cl co-transporter NKCC2. Nat Genet 1996;13:183–188.
8 Simon DB, Karet FE, Rodriguez-Soriano J, Hamdan JH, DiPietro A, Trachtman H, Sanjad SA, Lifton RP: Genetic heterogeneity of Bartter's syndrome revealed by mutations in the K^+ channel, ROMK. Nat Genet 1996;14: 152–156.
9 International Collaborative Study Group for Bartter-Like Syndromes: Mutations in the gene encoding the inwardly-rectifying renal potassium channel, ROMK, cause the antenatal variant of Bartter syntrome: Evidence for genetic heterogeneity. Hum Mol Genet 1997;6: 17–26.

10 Lloyd SE, Pearce SHS, Gunter W, Kawaguchi H, Igarashi T, Jentsch TJ, Thakker RV: Idiopathic low molecular weight proteinuria associated with hypercalciuric nephrocalcinosis in Japanese children is due to mutations of the renal chloride channel (CLCN5). J Clin Invest 1997;99:967–974.
11 Lloyd SE, Gunter W, Pearce SHS, Thomson A, Bianchi ML, Bosio M, Craig IW, Fisher SE, Scheinman SJ, Wrong O, Jentsch TJ, Thakker RV: Characterisation of renal chloride channel, CLCN5, mutations in hypercalciuric nephrolithiasis (kidney stones) disorders. Hum Mol Genet 1997;6:1233–1239.
12 Akuta N, Lloyd SE, Igarashi T, Shiraga H, Matsuyama T, Yokoro S, Cox JPD, Thakker RV: Mutational analysis of the renal chloride channel (CLCN5) in Japanese children with idiopathic low molecular weight proteinuria, hypercalciuria and nephrocalcinosis. Kidney Int 1997;52:911–916.
13 Wrong OM, Norden AGW, Feest TG: Dent's disease: A familial proximal renal tubular syndrome with low-molecular-weight proteinuria, hypercalciuria, nephrocalcinosis, metabolic bone disease, progressive renal failure and a marked male predominance. QJM 1994;87: 473–493.
14 Pook MA, Wrong OM, Wooding C, Norden AGW, Feest TG, Thakker RV: Dent's disease, a renal Fanconi syndrome with nephrocalcinosis and kidney stones, is associated with a micro-deletion involving DXS255 and maps to Xp11.22. Hum Mol Genet 1993;2:2129–2134.
15 Frymoyer PA, Scheinman SJ, Dunham PB, Jones DB, Hueber P, Schroeder ET: X-linked recessive nephrolithiasis with renal failure. N Engl J Med 1991;325:681–686.
16 Scheinman SJ, Pook MA, Wooding C, Pang JT, Frymoyer PA, Thakker RV: Mapping the gene causing X-linked recessive nephrolithiasis to Xp11.22 by linkage studies. J Clin Invest 1993; 91:2351–2357.
17 Bolino A, Devoto M, Enia G, Zoccali C, Weissenbach, Romeo G: Genetic mapping in the Xp11.2 region of a new form of X-linked hypophosphatemic rickets. Eur J Hum Genet 1993; 1:269–279.
18 Igarashi T, Hayakawa H, Shiraga H, Kawato H, Yan K, Kawaguchi H, Yamanaka T, Tsuchida S, Akagi K: Hypercalciuria nephracalcinosis in patients with idiopathic low molecular weight proteinuria in Japan: Is this identical to Dent's disease in the United Kingdom? Nephron 1995;69:242–247.
19 Oudet C, Martin-Coignard D, Pannetier S, Praud E, Champion G, Hanauer A: A second family with XLRH displays the mutation S244L in the CLCN5 gene. Hum Genet 1997; 99:781–784.
20 Nakazato H, Hattori S, Furuse A, Kawano T, Karashima S, Tsuruta M, Yoshimutao T, Endo F, Matsuda I: Mutations in the CLCN5 gene in Japanese patients with familial idiopathic low-molecular-weight proteinuria. Kidney Int 1997;52:895–900.
21 Fisher SE, Black GCM, Lloyd SE, Wrong OM, Thakker RV, Craig IW: Isolation and partial characterisation of a human chloride channel gene which is expressed in kidney and is a candidate for Dent's disease (an hereditary nephrolithiasis). Hum Mol Genet 1994;3:2053–2059.
22 Fisher SE, van Bakel I, Lloyd SE, Pearce SHS, Thakker RV, Craig IW: Cloning and characterisation of ClC-5, the human kidney chloride channel gene implicated in Dent's disease (an X-linked hereditary nephrolithiasis). Genomics 1995;29:598–606.
23 Kelleher CL, Buckalew OM, Frederickson ED, Rhodes DJ, Conner DA, SeidmanJG, Seidman CE: CLCN5 mutation Ser244Leu is associated with X-linked renal failure without X-linked recessive hypophosphataemic rickets. Kidney Int 1998;53:31–37.
24 Wrong OM, Unwin R, Cohen E, Tanner M, Thakker RV: Unravelling of the molecular mechanisms of kidney stones. Lancet 1996; 348:1561–1565.
25 Scheinman SJ: X-linked hypercalciuric nephrolithiasis. Clinical syndromes and chloride channel mutations. Kidney Int 1998;53:3–17.
26 Grenningloh G, Rienitz A, Schmitt B, Methfessel C, Zensen M, Beyreuther K, Gundelfinger D, Betz H: The strychnine-binding sub-unit of the glycine receptor shows homology with nicotinic acetylcholine receptors. Nature 1987;328: 215–220.
27 Schofield JR, Darlison MG, Fujita H, Burt DR, Stephenson FA, Rodriguez H, Rhee LM, Ramachandran J, Reale V, Glencorse TA, Seeburg PH, Barnard EA: Sequence and functional expression of the $GABA_A$ receptor shows a ligand-gated receptor super-family. Nature 1987;328:221–227.
28 Riordan JR, Rommens JM, Kerem B, Alon N, Rozmahel R, Grzelczak Z, Zielenski J, Lok S, Plavsic N, Chou JL, Drumm ML, Iannuzzi MC, Collins FS, Tsui LC: Identification of the cystic fibrosis gene: Cloning and characterization of complementary DNA. Science 1989; 245:1066–1073.
29 Jentsch TJ, Steinmeyer K, Schwarz G: Primary structure of *Torpedo marmorata* chloride channel isolated by expression cloning in *Xenopus* oocytes. Nature 1990;348:510–514.
30 Jentsch TJ, Gunther W, Pusch M, Schwappach B: Properties of voltage-gated chloride channels of the ClC gene family. J Physiol 1990;482: 19S–25S.
31 Fahlke C, Yu HT, Beck CL, Rhodes TH, George AL JR: Pore-forming segments in voltage-gated chloride channels. Nature 1997; 390:529–532.
32 Schmidt-Rose T, Jentsch TJ: Transmembrane topology of a CLC chloride channel. Proc Natl Acad Sci USA 1997;94:7633–7638.
33 Brandt S, Jentsch TJ: CL-6 and CLC-7 are two novel broadly expressed members of the CLC chloride channel family. FEBS Lett 1995;377: 15–20.
34 Ludewig U, Pusch M, Jentsch TJ: Two physically distinct pores in the dimeric CLC-0 chloride channel. Nature 1996;383:340–343.
35 Middleton RE, Pheasant DJ, Miller C: Homodimeric architecture of a CLC-type chloride ion channel. Nature 1996;383:337–340.
36 Koch MC, Steinmeyer K, Lorenz C, Ricker K, Wolf F, Otto M, Zoll B, Lehmann-Horn F, Grzeschik K-H, Jentsch TJ: The skeletal muscle chloride channel in dominant and recessive human myotonia. Science 1992;257:797–800.
37 Steinmeyer K, Lorenz C, Pusch M, Koch MC, Jentsch TJ: Multimeric structure of CLC-l chloride channel revealed by mutations in dominant myotonia congenita (Thomsen). EMBO J 1994;13:737–743.
38 Meyer-Kleine C, Steinmeyer K, Ricker K, Jentsch TJ, Koch-MC: Spectrum of mutations in the major human skeletal muscle chloride channel gene (CLCN1) leading to myotonia. Am J Hum Genet 1995;57:1325–1334.
39 Matsumura Y, Uchida S, Kondo Y, Miyazaki H, Kos BH, Hayama A, Morimoto T, Liu W, Arisalva M, Sasaki S, Maruma F: Overt nephrogenic diabetes insipidus in mice lacking the CLC-K1 chloride channel. Nat Genetics 1999; 21:95–98.
40 Cox JPD, Yamamoto K, Christie PT, Wooding C, Feest T, Flinter FA, Goodyer PR, Leumann E, Neuhaus T, Reid C, Williams PF, Wrong O, Thakker RV: Renal chloride channel, CLCN5, mutations in Dent's disease. J Bone Miner Res 1999;14:1536–1542.
41 Yamamoto K, Cox JPDT, Friedrich T, Christie PT, Bald M, Houtman PN, Marta J, Lapsley MJ, Patzer L, Picon G, Van't Hoff WG, Yamaoka K, Jentsch TJ, Thakker RV: Characterisation of renal chloride channel (CLCN5) mutations in Dent's disease. J Am Soc Nephrol, in press.
42 Devuyst O, Christie PT, Courtoy PJ, Beaumwens R, Thakker RV: Intra-renal and subcellular distribution of the human chloride channel, CLC-5, reveals a pathophysiological basis for Dent's disease. Hum Mol Genet 1999;8:247–257.
43 Gunther W, Luchow A, Cluzeaud F, Vandewalle A, Jentsch TJ: C1C-5, the chloride channel mutated in Dent's disease, colocalizes with the proton pump in endocytotically active kidney cells. Proc Natl Acad Sci USA 1998;95: 8075–8080.
44 Friedrich T, Breiderhoff T, Jentsch TJ: Mutational analysis demonstrates that C1C-4 and C1C-5 directly mediate plasma membrane currents. J Biol Chem 1999;274:896–902.
45 Hebert SC: Crystal clear chloride channels. Nature 1996;379:398–399.
46 Christensen E, Nielsen S: Structural and functional features of protein handling in the kidney proximal tubule. Semin Nephrol 1991;11: 414–439.
47 Nykjaer A, Dragun D, Walther D, Vorum H, Jacobsen C. Herz J, Melsen F, Christensen EI, Willnow TE: An endocytic pathway essential for renal uptake and activation of the steroid 25-(OH) vitamin D_3. Cell 1996;507–515.

48 Luyckx VA, Goda FO, Mount DB, Nishio T, Hall A, Hebert SC, Hammond DG, Yu AS: Intrarenal and subcellular localization of rat CLC5. Am J Physiol 1998;275:F761–F769.
49 Reinhart SC, Norden AG, Lapsley M, Thakker RV, Pang J, Moses AM, Frymoyer PA, Favus MJ, Hoepner JA, Scheinman SJ: Characterization of carrier females and affected males with X-linked recessive nephrolithiasis. J Am Soc Nephrol 1995;5:1451–1461.
50 Bateman A: The structure of a domain common to archaebacteria and the homocystinuria diseae protein. Trends Biochem Sci 1997;22:12–13.
51 Ponting CP: CBS domains in CLC chloride channels implicated in myotonia and nephrolithiasis (kidney stones). J Mol Med 1997;75:160–163.
52 Jentsch TJ, Friedrich T, Schriever A, Yamada H: The CLC chloride channel family. Pflügers Arch 1999;437:783–795.
53 Schwappach B, Stobrawa S, Hechenberger M, Steinmeyer K, Jentsch TJ: Golgi localization and functionally important domains in the NH2 and COOH terminus of the yeast CLC putative chloride channel Gef1p. J Biol Chem 1998;273:15110–15118.
54 Sintchak MD, Fleming MA, Futer O, Raybuck SA, Chambers SP, Caron PR, Murcko MA, Wilson KP: Structure and mechanism of inosine monophosphate dehydrogenase in complex with the immunosuppressant mycophenolic acid. Cell 1996;85:921–930.
55 Celenza JL, Eng FJ, Carlson M: Molecular analysis of the SNF4 gene of *Saccharomyces cerevisiae:* Evidence for physical association of the SNF4 protein with the SNF1 protein kinase. Mol Cell Biol 1989;9:5045.
56 Hoopes RR Jr, Hueber PA, Reid RJ, Braden GL, Goodyer PR, Melnyk AR, Midgley JP, Moel DI, Neu AM, Van Why SK, Scheinman SJ: CLCN5 chloride channel mutations in six new North American families with X-linked nephrolithiasis. Kidney Int 1998;54:698–705.
57 Morimoto T, Uchida S, Sakamoto H, Kondo Y, Hanamizu H, Fukui M, Tomino Y, Nagano N, Sasaki S, Marumo F: Mutations in CLCN5 chloride channel in Japanese patients with low molecular weight proteinuria. J Am Soc Nephrol 1998;9:811–818.
58 Igarashi T, Gunther W, Sekine T, Inatomi J, Shiraga H, Takahashi S, Suzuki J, Tsuru N, Yanagihara T, Shimazu M, Jentsch TJ, Thakker RV: Functional characterization of renal chloride channel, CLCN5, mutations associated with Dent's Japan disease. Kidney Int 1998;54:1850–1856.
59 Igarashi T, Iantomi J, Ohara T, Kuwahara T, Shimadzu M, Thakker RV: Clinical and genetic studies of CLCN5 mutations in Japanese families with Dent's disease. Kidney Int, in press.

Experimental Nephrology

Exp Nephrol 2000;8:361–365

Severely Impaired Urine-Concentrating Ability in Mice Lacking the CLC-K1 Chloride Channel

Shinichi Uchida   Fumiaki Marumo

Second Department of Internal Medicine, Tokyo Medical and Dental University, School of Medicine, Tokyo, Japan

Key Words

Thin limb of Henle's loop · Gene targeting · Countercurrent system · Nephrogenic diabetes insipidus, mice

Abstract

To analyze the physiological functions of CLC-K1 in vivo, we generated mice lacking CLC-K1 by targeted gene disruption. Homozygous mutant Clcnk1–/– mice produced ~5 times more urine than Clcnk1+/– and Clcnk1+/+ mice. After 24-hour water deprivation, Clcnk1–/– mice became severely dehydrated and lethargic. Intraperitoneal injection of the V2 agonist, deamino-Cys^1, D-Arg^8 vasopressin, induced an increase in urine osmolarity in Clcnk1+/– and Clcnk1+/+ mice from ~1,000 to ~3,000 mosm/kg H_2O, whereas the increase in Clcnk1–/– mice was only from ~600 to ~840 mosm/kg H_2O, indicating nephrogenic diabetes insipidus in Clcnk1–/– mice. These results clearly established that CLC-K1 plays a major role in the urinary-concentrating mechanisms.

Introduction

CLC-K1 is a kidney-specific chloride channel exclusively present in the plasma membrane of the thin ascending limb of Henle's loop (tAL) in the inner medulla [1, 2]. In contrast to CLC-K2, whose physiological role has been established by identifying mutations in CLCKB genes (a human homolog of rat CLC-K2) in patients with Bartter's syndrome [3], the physiological role of CLC-K1 in vivo had not been clarified prior to this study. NaCl transport in the tAL is thought to be an important component of countercurrent systems for urine concentration, but there has been no direct evidence that CLC-K1 was indeed involved in the urinary-concentrating mechanisms. To investigate this issue, we have generated CLC-K1 knockout mice by targeted gene disruption and found that the CLC-K1–/– mice showed severely impaired urinary-concentrating ability.

In this review, we briefly summarize the characteristics of the CLC-K1 chloride channel and describe the phenotypes observed in CLC-K1–/– mice.

What Is CLC-K1?

Nine CLC chloride channels are now known to exist in mammals, and mutations in three of these human genes (CLCN1, CLCN5, and CLCNKB) have been shown to be associated with human diseases [3–5]. CLC-1, a skeletal muscle-specific CLC chloride channel, is responsible for the majority of the plasma membrane conductance of skeletal muscle, and its mutations lead to recessive and dominant types of myotonia [4]. Mutations of CLCN5, a gene which forms a subfamily with CLC-3 and CLC-4 [6,

1018–7782/00/0086–0361$17.50/0

Accessible online at:
www.karger.com/journals/exn

Shinichi Uchida
1-5-45 Yushima Bunkyo-ku
Tokyo, 113-8519 (Japan)
Fax +81 3 5803 0172, E-Mail suchida.med2@med.tmd.ac.jp

Table 1. Comparison between the CLC-K1 expressed in oocytes and a native chloride transport in the tAL

	CLC-K1	In vitro perfusion studies
Anion selectivity	Br > Cl >> I = gluconate	Br ≥ Cl >> I = gluconate
pH dependence	K_d = 6.9	K_d = 6.3
Ca sensitivity	Ca free in the bath solution reduced the expressed current to ~ 40%	Ca free in the bath solution reduced the chloride diffusion voltage and $^{36}Cl^-$ flux to ~ 40%.

7] and is relatively abundant in kidney, was recently found in several syndromes displaying nephrolithiasis such as Dent's disease, X-linked recessive hypophosphatemic rickets, and X-linked recessive nephrolithiasis [5]. Mutations of the CLCNKB gene were additionally found in patients with Bartter's syndrome [3]. CLC-KB is one of the two kidney-specific CLC chloride channels and a human homolog of rat CLC-K2 [8, 9], a gene expressed mainly in the thick ascending limb of Henle's loop, distal tubules, and collecting ducts [10]. Accordingly, CLC-K2 is speculated to serve as an exit for chloride ions on the basolateral surface of nephron segments which have Na^+-coupled Cl^- transporters on the apical side [11]. CLC-K1 is another member of the CLC-K channel family. Our group first isolated it as a kidney-specific CLC chloride channel by PCR-based homology cloning strategy in 1993 [1], and its precise localization in the kidney was later determined by immunohistochemistry [2]. In contrast to the relatively broad localization of CLC-K2 along the nephron [12], CLC-K1 is localized in the inner medulla [12], especially in the tAL [2]. Immunoelectron microscopy revealed that CLC-K1 is present in both the apical and basolateral plasma membranes of the tAL [2]. A chloride channel in both of these plasma membranes of a specific nephron segment makes it highly permeable to chloride. Accordingly, CLC-K1 could be a major channel responsible for the extraordinally high chloride permeability reported in the tAL [13]. To verify this, we compared the functional characteristics of CLC-K1 expressed in *Xenopus* oocytes [2] with a native chloride transport in the tAL [14–16]. As shown in table 1, both transport systems are quite similar. Anion selectivity sequences in both systems are Br > Cl >> I. CLC-K1 expressed in oocytes showed pH-sensitive chloride currents [2]. Hill's coefficient was 1.2, and K_d in the CLC-K1 current expressed in oocytes was 6.9, a value quite similar to the K_d of 6.3 measured in the in vitro perfusion study [14]. The effect of calcium removal from the bath solution also showed the same level of reduction of chloride transport or current. Although there were some functional differences in details between the perfusion studies and the oocyte expression study, we think that these two chloride transport systems have basically the same characteristics. Accordingly, together with the immunohistochemical data, we could conclude that CLC-K1 is a major chloride channel responsible for the high chloride permeability in the tAL.

Indirect Evidence that CLC-K1 May Be Important for Urinary Concentration

To embark on the knockout project of the CLC-K1 gene, we needed some evidence supporting the contention that CLC-K1 is indeed important for urinary concentration. One such piece of evidence is the functional regulation of chloride transport in the tAL by vasopressin. Takahashi et al. [17] reported an increase in chloride permeability in the tAL induced by vasopressin treatment. They concluded that this effect was mediated by the signal through the V2 receptor, suggesting that, when vasopressin secretion is increased, i.e., under a condition where the body needs a higher urine concentration, the chloride transport in the tAL should also increase, probably through CLC-K1. In other words, a higher chloride transport capacity in the tAL is required for a higher urinary concentration. The expression of CLC-K1 mRNA was also shown to be increased in dehydrated rats [1]. This finding also provides evidence that more CLC-K1 may be necessary for maximal urinary concentration.

Generation of the CLC-K1–/– Mice

To generate the CLC-K1–/– mice, we used homologous recombination in embryonic stem cells to create a mutant allele (Clcnk 1 minus) in which exons 1–8 were replaced by a cassette expressing the selective marker neomycin transferase [18]. Since the CLC-K2 gene is located near the CLC-K1 gene, a long arm of the targeting construct included the exons of CLC-K2. To verify the knock-

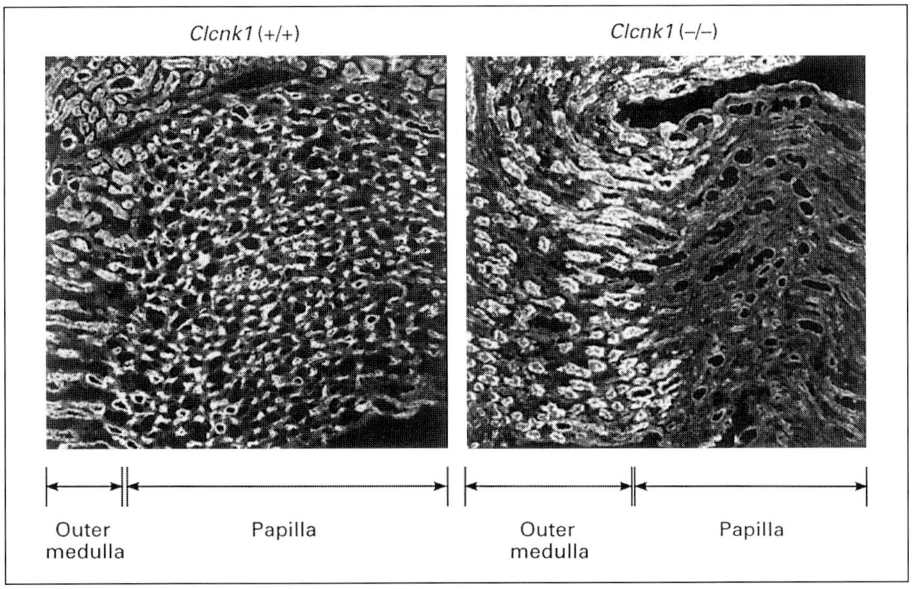

Fig. 1. Immunohistochemistry of mouse CLC-K1 and CLC-K2 in wild-type and CLC-K1 (–/–) mice. The staining in the outer medulla represents the CLC-K2, which was preserved in the CLC-K1(–/–) mice in the right panel.

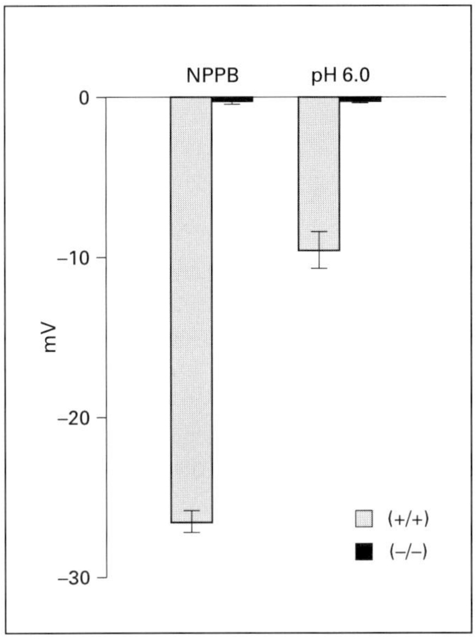

Fig. 2. In vitro perfusion study of the thin ascending limb. Transepithelial diffusion voltage of chloride was reduced by low pH (6.0) or a chloride channel blocker, 5-nitro-2-(3-phenyl-propylamino)-benzoate (NPPB), in the wild-type thin limb. However, no responses were observed in the CLC-K1 minus thin limb.

out of CLC-K1 and preservation of CLC-K2, we performed a ribonuclease protection assay [18] and immunohistochemistry. Ribonuclease protection assay probes were designed to cover the first 100 amino acids of mouse CLC-K1 and CLC-K2. Using these probes, we could confirm that the CLC-K1–/– mice did not express CLC-K1. Immunohistochemistry (fig. 1) was performed using the antibody previously prepared for rat CLC-K1. This antibody is specific for CLC-K1 in rats [2] but recognizes both mouse CLC-K1 and -K2. In the CLC-K1–/– mice, the staining in the tAL in the inner medulla was absent, whereas the staining in the thick ascending limb of Henle's loop and distal tubules was preserved (fig. 1). This result also clearly verified the selective disruption of the CLC-K1 gene in our knockout mice.

In vitro Microperfusion Study of the tAL from the CLC-K1–/– Mice

Before starting the analysis of the renal functions of CLC-K1–/– mice, we characterized the chloride transport in the tAL of the CLC-K1–/– mice to verify our hypothesis that the CLC-K1 is responsible for the high chloride permeability in the tAL. To observe the Cl^- permeability in the tAL, the transepithelial diffusional potential (V_d) of Cl^- was measured. When the bath Cl^- solution was decreased from 145 to 14.5 mmol/l by substituting chloride with gluconate, the steady-state V_ds were 36.7 ± 0.7 (n = 8) and 6.2 ± 1.1 mV (n = 8) in wild-type and knockout mice, respectively. When the bath pH was decreased from 7.4 to 6.0, the decreases in V_ds were 9.7 ± 1.2 (n = 8) and 0.3 ± 0.1 mV (n = 8) in wild-type and knockout mice, respectively. When 0.1 mmol/l 5-nitro-2-(3-phenyl-propylamino)-benzoate, a specific Cl^- channel blocker, was applied to the bath, the V_ds were changed by –26.7 ± 0.6 (n = 8) and –0.1 ± 0.2 mV (n = 8) in wild-type and knockout mice, respectively (fig. 2). These observations clearly confirmed that the previously identified chloride transport activity in the tAL is mediated by the CLC-K1 chloride channel [14, 15, 19, 20].

Plasma and Urine Data of the CLC-K1–/– Mice

Once we confirmed the gene knockout of CLC-K1 at the mRNA, protein, and functional levels, we embarked on an analysis of the renal function of the CLC-K1–/–

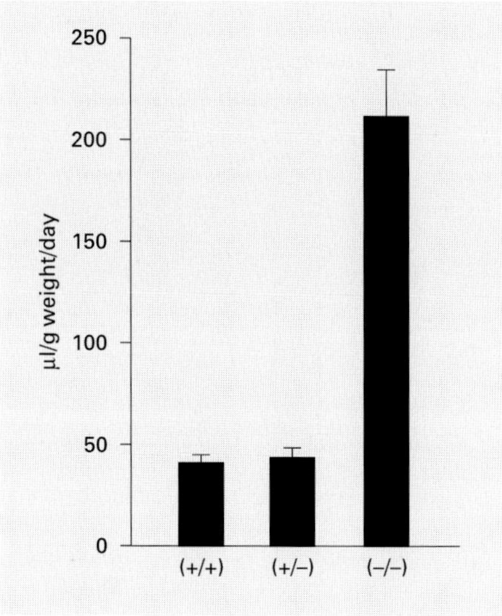

Fig. 3. Urine volume of euhydrated wild-type, heterozygous, and knockout mice.

mice. The CLC-K1–/– mice were not obviously different from wild-type mice in survival, gross physical appearance, or organ morphology. To determine the physiological role of CLC-K1 in vivo, urine and plasma samples were collected under normal and dehydrated conditions in metabolic cages. Analysis of plasma from wild-type, heterozygous, and knockout mice revealed no significant differences in pH values and the concentrations of Na^+, K^+, Cl^-, and HCO_3^- among the 3 groups. This could be taken as a clear indication that, unlike CLC-KB mutations the loss-of-function mutation of CLC-K1 does not result in hypokalemic alkalosis (Bartter's syndrome). The plasma creatinine level in the knockout mice was normal, indicating that the glomerular filtration rate was not affected in these animals. In contrast to these plasma parameters, the CLC-K1–/– mice showed a remarkably increased urine output (~5 times more urine compared with wild-type and heterozygous mice; fig. 3). Urinary excretions of sodium, potassium, and chloride were not significantly different among the 3 groups.

Effect of Dehydration

After 24 h of water deprivation, the CLC-K1–/– mice became severely dehydrated and lethargic, their body weights decreased by 27% (10% decrease in the heterozygous and wild-type mice), and their serum osmolarity increased to > 500 mosm/kg H_2O (310 mosm/kg H_2O in the heterozygous and wild-type mice). Although urine osmolarity in the heterozygous and wild-type mice was increased from 1,000 to 3,000 mosm/kg H_2O after intraperitoneal injection of the V2 agonist deamino-Cys1, D-Arg8 vasopressin, that in the CLC-K1–/– mice was increased only slightly from 600 to 800 mosm/kg H_2O, indicating the latter's inability to generate hypertonic medullary interstitium.

Lessons from the CLC-K1–/– Mice

Renal urinary concentrating mechanisms consist of two important processes. The first one is the osmotic equilibration of collecting duct fluid with hypertonic medullary interstitial fluid, a process which is regulated by vasopressin. The loss-of-function mutations in the vasopressin V2 receptor and the AQP-2 water channel in patients with nephrogenic diabetes insipidus clearly demonstrated the importance of this process. The second is the generation and maintenance of hypertonic medullary interstitium as a result of the operation of a countercurrent system. Imai et al. [21] and Kokko and Rector [22] have proposed passive models of countercurrent multiplication systems in the inner medulla of the kidney in which the extraordinally high chloride permeability in the tAL was thought to be an important component. However, there has been no direct evidence that CLC-K1 is a chloride channel responsible for the high chloride permeability in the tAL or that a chloride transport through CLC-K1 is important for the countercurrent multiplication system. The data above clearly demonstrated for the first time that CLC-K1 is important in the renal urinary-concentrating mechanisms and that the countercurrent system is indeed involved in urinary-concentrating mechanisms. Findings on these animals also suggested the possibility that mutations of the CLC-K1 gene in human (CLCNKA) may be associated with nephrogenic diabetes insipidus having no mutations in the vasopressin V2 receptor and AQP-2 water channel genes. So far, mutations of CLCNKA have not been reported in patients with nephrogenic diabetes insipidus. Considering that AQP-1–/– mice also showed severely impaired urinary-concentrating ability [23] while patients with the AQP-1 mutation did not show overt diabetes insipidus [24], it is possible that the urine-concentrating defect in CLC-K1–/– is less severe in humans that in mice. The reason for this species difference is not fully understood. Specific constituents of the countercurrent systems may differ between mice and humans, and the difference in the composition of osmolytes in the inner medulla, i.e., the ratio of NaCl versus urea [25], may support this possibility. In *Psammomys,* the content of NaCl is an order of magnitude greater than that of urea under maximal urine concentration. On the other hand, urea contributes about three times as much as NaCl to papillary solutes in the rat. Accordingly, it is speculated that the defect of NaCl transport in the tAL may have a more significant effect on the accumulation of NaCl than the accumulation of urea, thus resulting in a more severe urine-concentrating defect in mouse species in which NaCl may be the dominant papillary solute. We are currently measuring the solute content in the papillae of CLC-K1–/– mice to ascertain wether the defect in CLC-K1 impairs urea accumulation as well as NaCl.

References

1 Uchida S, Sasaki S, Furukawa T, Hiraoka M, Imai T, Hirata Y, Marumo F: Molecular cloning of a chloride channel that is regulated by dehydration and expressed predominantly in kidney medulla. J Biol Chem 1993;268:3821–3824.

2 Uchida S, Sasaki S, Nitta K, Uchida K, Horita S, Nihei H, Marumo F: Localization and functional characterization of rat kidney-specific chloride channel, CIC-K1. J Clin Invest 1995; 95:104–113.

3 Simon DB, Bindra RS, Mansfield TA, Nelson-Williams C, Mendonca E, Stone R, Schurman S, Nayir A, Alpay H, Bakkaloglu A, Rodriguez-Soriano J, Morales JM, Sanjad SA, Taylor CM, Pilz D, Brem A, Trachtman H, Griswold W, Richard GA, John E, Lifton RP: Mutations in the chloride channel gene, *CLCNKB,* cause Bartter's syndrome type III. Nat Genet 1997; 17:171–178.

4 Koch MC, Steinmeyer K, Lorenz C, Ricker K, Wolf F, Otto M, Zoll B, Lehmann-Horn F, Grzeschik K-H, Jentsch TJ: The skeletal muscle chloride channel in dominant and recessive human myotonia. Science 1992;257:797–800.

5 Lloyd SE, Pearce SHS, Fisher SE, Steinmeyer K, Schwappach B, Scheinman SJ, Harding B, Bolino A, Devoto M, Goodyer P, Rigden SPA, Wrong O, Jentsch TJ, Craig IW, Thakker RV: A common molecular basis for three inherited kidney stone diseases. Nature 1996;379:445–449.

6 Kawasaki M, Uchida S, Monkawa T, Miyawaki A, Mikoshiba K, Marumo F, Sasaki S: Cloning and expression of a protein kinase C-regulated chloride channel abundantly expressed in rat brain neuronal cells. Neuron 1994;12:597–604.

7 Jentsch TJ, Günther W, Pusch M, Schwappach B: Properties of voltage-gated chloride channels of the ClC gene family. J Physiol (Lond) 1995;482:19S–25S.

8 Kieferle S, Fong P, Bens M, Vandewalle A, Jentsch TJ: Two highly homologous members of the ClC chloride channel family in both rat and human kidney. Proc Natl Acad Sci USA 1994;91:6943–6947.

9 Takeuchi Y, Uchida S, Marumo F, Sasaki S: Cloning, tissue distribution, and intrarenal localization of ClC chloride channels in human kidney. Kidney Int 1995;48:1497–1503.

10 Adachi S, Uchida S, Ito H, Hata M, Hiroe M, Marumo F, Sasaki S: Two isoforms of a chloride channel predominantly expressed in thick ascending limb of Henle's loop and collecting ducts of rat kidney. J Biol Chem 1994;269: 17677–17683.

11 Gamba G, Miyanoshita A, Lombardi M, Lytton J, Lee W-S, Hediger MA, Hebert SC: Molecular cloning, primary structure, and characterization of two members of the mammalian electroneutral sodium-(potassium)-chloride cotransporter family expressed in kidney. J Biol Chem 1994;269:17713–17722.

12 Yoshikawa M, Uchida S, Yamauchi A, Miyai A, Tanaka Y, Sasaki S, Marumo F: Localization of rat CLC-K2 chloride channel mRNA in the kidney. Am J Physiol 1999;276:F552–F558.

13 Imai M, Kokko JP: Sodium, chloride, urea, and water transport in the thin ascending limb of Henle. Generation of osmotic gradients by passive diffusion of solutes. J Clin Invest 1974;53: 393–402.

14 Kondo Y, Yoshitomi K, Imai M: Effect of pH on Cl^- transport in the thin limb ascending limb of Henle's loop. Am J Physiol 1987;253: F1216–F1222.

15 Kondo Y, Yoshitomi K, Imai M: Effect of Ca^{2+} on Cl^- transport in the thin ascending limb of Henle's loop. Am J Physiol 1988;254:F232–F239.

16 Isozaki T, Yoshitomi K, Imai M: Selectivity of ion permeability across ascending thin limb of Henle's loop: Interaction of Cl^- and other halogens with anion transport system. Kidney Int Suppl 1991;40:S113–S118.

17 Takahashi N, Kondo Y, Ito O, Igarashi Y, Omata K, Abe K: Vasopressin stimulates Cl^- transport in ascending thin limb of Henle's loop in hamster. J Clin Invest 1995;95:1623–1627.

18 Matsumura Y, Uchida S, Kondo Y, Miyazaki H, Ko SBH, Hayama A, Morimoto T, Liu W, Arisawa M, Sasaki S, Marumo F: Overt nephrogenic diabetes insipidus in mice lacking the CLC-K1 chloride channel. Nat Genet 1999;21: 95–98.

19 Kondo Y, Yoshitomi K, Imai M: Effect of anion transport inhibitors and ion substitution on Cl transport in the thin ascending limb of Henle's loop. Am J Physiol 1987;253:F1206–F1215.

20 Yoshitomi K, Kondo Y, Imai M: Evidence for conductive Cl pathways across the cell membranes of the thin ascending limb of Henle's loop. J Clin Invest 1988;82:866–871.

21 Imai M, Taniguchi J, Tabei K: Function of thin loop of Henle. Kidney Int 1987;31:565–579.

22 Kokko JP, Rector FC: Countercurrent multiplication system without active transport in inner medulla. Kidney Int 1972;2:214–223.

23 Ma T, Yang B, Gillespie A, Carlson EJ, Epstein CJ, Verkman AS: Severely impaired urinary concentrating ability in transgenic mice lacking Aquaporin-1 water channel. J Biol Chem 1997; 273:4296–4299.

24 Preston GM, Smith BL, Zeidel ML, Moulds JJ, Agre P: Mutations in *aquaporin-1* in phenotypically normal humans without functional CHIP water channels. Science 1994;265:1585–1587.

25 Jamison RL, Roy DR, Layton HE: Countercurrent mechanism and its regulation; in Seldin DW, Giebisch G (eds): Clinical Disturbances of Water Metabolism. New York, Raven Press, 1993, pp 119–156.

Experimental Nephrology

Exp Nephrol 2000;8:366–375

Molecular Characteristics of Phosphate Transporters and Their Regulation

Nati Hernando Ian C. Forster Jürg Biber Heini Murer

Institute of Physiology, University of Zurich, Switzerland

Key Words
Proximal tubule · P_i transport · Protein trafficking · Parathyroid hormone

Abstract
A key process in overall P_i-homeostasis is renal proximal tubular reabsorption of inorganic phosphate (P_i), which involves secondary active sodium/phosphate (Na^+/P_i) cotransport reabsorption at the brush border membrane. Among the two different molecularly identified Na^+/P_i cotransporters, the type-IIa Na^+/P_i cotransporter (NaPi-IIa) accounts for up to 70% of brush border membrane transport. Regulation of renal P_i reabsorption centers around brush border membrane insertion and retrieval of transporter protein under the influence of hormonal and nonhormonal factors. Immunohistochemical and fluorescence techniques have provided new insights into the tissue distribution and the regulation processes. The intrinsic electrogenicity of NaPi-IIa, has allowed detailed studies of the transport kinetics of NaPi-IIa and, combined with mutagenesis methods, structure-function information at the protein level is emerging.

Introduction

Inorganic phosophate (P_i) is an essential element in cellular metabolism, being required for the formation of bone, phospholipids, nucleotides and nucleic acids, and for intracellular signal transduction. Therefore, P_i homeostasis must be tightly controlled. This control results from the balance between intestinal absorption and renal excretion. The latter is the result of free glomerular filtration and reabsorption in the proximal tubule, where up to 80% of filtered P_i is reabsorbed by a Na^+-coupled, secondary active transport mechanism. P_i entry at the brush border membrane involves Na^+/P_i cotransporters and these are the final target for physiological and pathophysiological regulation of renal P_i reabsorption. The basolateral exit pathways are ill-defined.

To date three different Na^+/P_i-cotransporters have been identified that have been classified as type I, type II and type III [1]. The type-I transporter is found in renal proximal tubular brush border membranes and mediates, in addition to Na^+/P_i cotransport, a multi-specific anion conductance. The type-II family can be divided in two subgroups, type IIa and type IIb, which show an overall homology of about 60%. The major differences between both groups are the amino acid sequences in the carboxy terminus of each protein [2] and their tissue distribution.

KARGER
Fax + 41 61 306 12 34
E-Mail karger@karger.ch
www.karger.com

1018–7782/00/0086–0366$17.50/0

Accessible online at:
www.karger.com/journals/exn

Nati Hernando, PhD/Ian Forster, PhD
Institute of Physiology, University of Zurich
Winterthurerstrasse 190, CH–8057 Zurich (Switzerland)
Tel. +41 1 635 5030, Fax +41 1 635 5715
E-Mail hernando@physiol.unizh.ch/forster@physiol.unizh.ch

Type IIa is expressed in the apical membrane of the renal proximal epithelium [3], whereas type IIb, which is not detected in mammalian renal tissues, is located on the apical membrane of the intestinal and lung epithelia [2, 4]. Finally, the retroviral receptors Glvr-1 and Ram-1, which are expressed both in epithelial and non-epithelial cells, have been designated as type III [5]. From these three main groups of Na^+/P_i cotransporters, the type-IIa subgroup is chiefly responsible for reabsorption of P_i in the kidney, as suggested by its functional properties and specific tissue distribution. Moreover, this role is also supported by antisense experiments performed both in vivo and in vitro [6, 7], as well as by the renal P_i reabsorption characteristics of a mouse knock-out model [8].

In this review we shall focus on the renal type-IIa system (NaPi-IIa) because of its physiological importance, and review the current knowledge concerning molecular characterization, physiological regulation at the molecular level, transport kinetics and structure-function properties. The reader is referred to previous reviews which cover other aspects of Na^+-coupled P_i cotransport [1, 9–13].

Molecular Characterization of the Renal-Specific Type-IIa Na/P_i Cotransporter

The type IIa Na^+/P_i cotransporter was first identified at the molecular level by expression cloning in *Xenopus laevis* oocytes injected with cRNAs from rabbit and human kidneys. Later on, the related genes of a number of mammals were cloned by probe hybridization.

Gene Characterization

The gene encoding the type-IIa Na/P_i cotransporter (NPT2) has been cloned from several species [14–16]. The human and murine genes contain 13 exons and 12 introns, with donor and acceptor consensus sequences in their junctions [14, 15]. Putative CAAT and TATA boxes are located respectively at positions –143/–147 and –51/–40 relative to the first nucleotide of the cDNAs. In both cases the translation initiation site is located within exon 2. No correlation was detected between the predicted secondary structure model and the exon-defined peptide domains. For the murine gene, two transcription initiation and two polyadenylation sites have been identified. The NPT2 promoter was functional in cells from renal proximal origin (OK), but not from distal tubules (MDCK) [17]. The promoter activity is regulated by CO_2/HCO_3^{2-} and vitamin D_3 [18, 19]; transcriptional regulation has also been proposed for thyroid hormone (T_3) [20]. However, other factors that are known to modulate the activity of the cotransporter in vivo, such as parathyroid hormone or medium P_i content, do not have an effect on the promoter activity [16].

mRNA Expression

Expression of mRNA is restricted to the proximal tubules, preferentially in deep nephrons [3, 21]. Northern blots of RNA samples from different species detected a single major band at around 2.7 kb only in the kidney. A recent study that reported the presence of four transcripts of 9.5, 4.6, 2.6 and 1.2 kb in the rat kidney needs further confirmation [22]. RT-PCR performed with single microdissected rat nephron segments showed that the mRNA is expressed in proximal convoluted and proximal straight tubules, but not in glomeruli, thick ascending limb of Henle's loop or collecting ducts [3]. The mRNA levels remain mostly unaffected by treatment with factors known to regulate renal P_i reabsorption. With the exception of T_3 and vitamin D_3, when changes in mRNA levels were reported, they seem to reflect changes in mRNA stability, rather than modification of the transcription rate, which is in agreement with the regulation of the promoter activity in vitro.

Protein Characterization

The NPT2 gene encodes a protein of about 640 amino acids, with a predicted molecular mass of ~70 kD. However, in Western blots performed under nonreducing conditions, the NaPi-IIa protein appears as a band with an apparent mass of 95–120 kD, due to heavy glycosylation [23]. Under reducing conditions two bands of 45–50 kD appear, suggesting a proteolytic cleavage [24, 25]. Hydropathy plots predict eight transmembrane domains, with both the amino and carboxy tails facing the cytoplasm (fig. 1). This model has been partially confirmed by selective introduction of FLAG epitopes as well as by cysteine scanning experiments [26, 27] (fig. 1). Several potential N-glycosylation sites within the second extracellular loop are used, as suggested by the higher than expected apparent mass. Based on site-directed mutagenesis experiments, the glycosylation does not seem essential for activity, although the surface delivery of the mutated cotransporters was slightly reduced in oocytes [28]. In agreement with the mRNA data, protein expression is highest in the convoluted part of proximal tubules and gradually decreases along the straight part [3, 21]. This distribution fully agrees with the sites of renal P_i reabsorption.

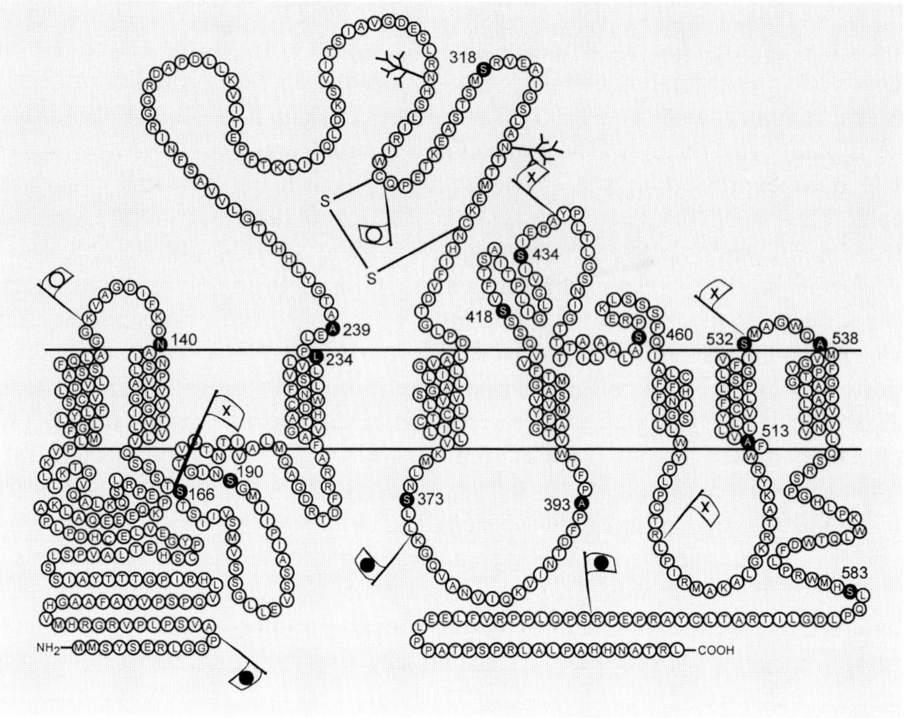

Fig. 1. Molecular model for NaPi-IIa. A secondary structure of NaPi-IIa based on hydrophobicity data predicts 8 transmembrane spanning domains. From a FLAG epitope study [26], it was concluded that the amino and carboxy termini and the region connecting transmembrane domains TM 4 and 5 are located intracellularly, whereas the loops connecting TM 1 and 2, and TM 3 and 4 are extracellular. Flags indicate insertion points for the FLAG epitopes: flags with circles indicate constructs that produce functional transport (open = extracellular; filled = intracellular location); flags with crosses indicate constructs that prevented protein from functioning. In a cysteine scanning study [27] amino acids indicated by filled circles were replaced by cysteines and tested for accessibility by impermeant alkylating reagents. Only the transport function of mutant S460C was affected by alkylation, which indicated that this is located in a functionally important region.

Polarized Distribution of NaPi-IIa

Within the tubules, NaPi-IIa is specifically located in the apical (brush border) membrane. This asymmetric distribution can be reproduced in cells in culture, but it requires cells having a renal proximal tubular origin. Thus, specific apical targeting NaPi-IIa was achieved in OK cells (derived from opossum renal proximal cells) transfected with rat NaPi-IIa, but not im MDCK cells (a canine distal tubule cell line) or CaCo-2 cells (derived from human intestine) [29]. Partial truncations of the carboxy terminal tail indicate that it contains sequences required for proper sorting [30]. In contrast to NaPi-IIa, the mouse isoform of the intestinal P_i cotransporter (NaPi-IIb) appears to be apically sorted in all the cellular models analyzed so far [29]. This difference in membrane delivery offers the possibility to identify sorting signals by studying the behavior of type IIa-type IIb chimeras.

The brush border membrane expression of NaPi-IIa is regulated by all factors or conditions known to control P_i homeostasis. Thus, the NaPi-IIa is downregulated in situations associated with decreased renal P_i reabsorption, whereas it is upregulated in situations that induce an increase in renal P_i reabsorption. As indicated above, in most of the cases these changes take place without parallel alterations in mRNA levels, suggesting that they involve regulation of the protein stability.

Downregulation of NaPi-IIa

Parathyroid Hormone Treatment

The phosphaturic effect associated with parathyroid hormone (PTH) treatment is due to a reduction in renal P_i reabsorption. Analysis of renal proximal tubules (fig. 2) as well as brush border preparations from rats treated with PTH showed that a rapid downregulation of NaPi-IIa mediates the renal response [31]. This effect can be reproduced in OK cells in culture, where it has been analyzed in more detail. Based on in vivo and in vitro experiments, it is now clear that PTH induces the endocytosis and lysosomal degradation of NaPi-IIa [32–34]. This process occurs in two steps: internalization of the cotransporter into a subapical tubulovesicular compartment (independent of microtubules), and its delivery to lysosomes for degradation (dependent on microtubules) [35]. Although it cannot be ruled out completely, endocytosed cotransporters do not seem to recycle back to the plasma membrane. This concept is supported by the finding that recovery of basal cotransporter levels after removal of PTH is totally dependent on de novo protein synthesis and on an intact microtubular network [33].

The PTH effect involves activation of the protein kinase A and protein kinase C (PKC) signalling pathways, but nothing is known about the subsequent steps. Al-

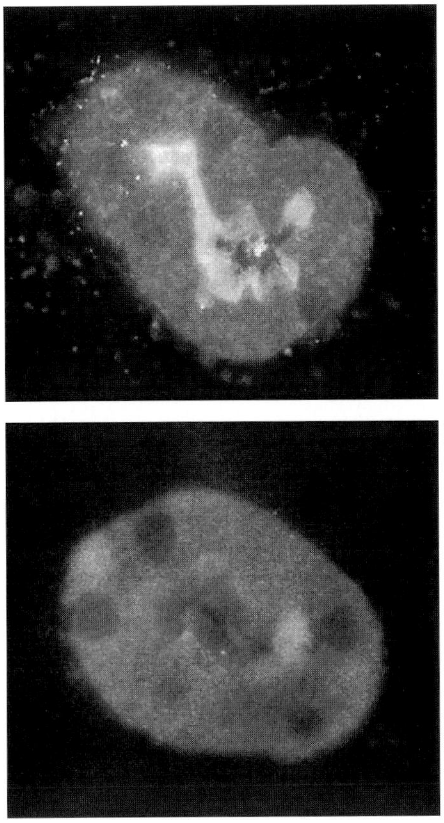

Fig. 2. Regulation by PTH. Isolated murine proximal tubules untreated (**a**) and treated (**b**) with 1–34 PTH. The preparations were stained with a specific type-IIa polyclonal antibody.

though the NaPi-IIa protein contains several potential PKC phosphorylation sites, they do not seem required for the PTH-induced endocytosis [36]. Activation of the PKC pathway in *X. laevis* oocytes that express NaPi-IIa, leads to loss of transport function that is directly correlated with internalization of NaPi-IIa protein, but no changes in the transport kinetics at the individual cotransporter level occur [37]. Moreover, coexpression of NaPi-IIa and the Na^+-sulfate cotransporter (NaSi-1) in *X. laevis* oocytes has shown that the PKC-induced downregulation is specific to Na-Pi-IIa protein [37, 38].

In general, the endocytic process involves at least two types of sorting signals located within cytoplasmic domains of the internalized protein: tyrosine and dileucine-based motifs [39]. In proximal tubules, the cotransporters colocalize in part with the adaptor protein AP2 in the intermicrovillar clefts, which suggests that endocytosis could take place in clathrin-coated vesicles [40]. Recent experiments indicate that the above-mentioned sorting signals do not play a role in the PTH-induced endocytosis of the type-IIa cotransporter [41].

High-Phosphate Diet

Renal P_i reabsorption is regulated by the P_i content of the diet. Administration of a diet containing high P_i levels induces a rapid endocytosis of NaPi-IIa that is independent of PTH [42]. A similar effect is observed after intravenous injection of phosphate, and it can also be reproduced in OK cells in culture by addition of media containing high P_i [43]. Its mechanism of action remains unclear, and it will be discussed in the next section together with the low P_i-induced upregulation of the cotransporter.

Glucocorticoids

Administration of glucocorticoids induces a PTH-independent inhibition of renal P_i reabsorption. This response is mediated by a decrease in the expression of NaPi-IIa. Changes in the lipid composition of the membrane as well as stimulation of unknown inhibitory proteins have been proposed as possible mechanisms [44].

Upregulation of NaPi-IIa

Developmental Upregulation

Renal development and functions are still immature at birth. In kidneys of newborn rats, expression of NaPi-IIa is restricted to differentiated juxtamedullary nephrons. Formation of a differentiated brush border during suckling is associated with the expression of the cotransporter protein and mRNA [45]. After nephrogenesis is complete, if the pups are still fed breast milk (low P_i content; see below), both superficial and juxtamedullary nephrons show high protein levels. After weaning, the pattern of expression is similar to that in adults, due to a downregulation of the cotransporter in superficial nephrons [45]. This last change may be related to both a higher P_i content of the food as well as upregulation of hormonal factors with phosphaturic activity, as glucocorticoids and PTH. A role of thyroid hormone in the maturation of the cotransporter has also been suggested [46]. Finally, the existence of a growth-specific cotransporter, related but different from the adult NaPi-IIa, has also been proposed [47].

Low-Phosphate Diet

As mentioned above, renal P_i reabsorption strongly correlates with the P_i intake. Administration of low P_i diet induces a rapid upregulation of NaPi-IIa that can be reproduced in vitro by addition of low P_i media to OK cells. Under acute conditions, there is a rapid increase in the number of cotransporters detected in the apical membrane of proximal tubules that is particularly evident in superficial and mid-cortical nephrons, where the expression of the cotransporter is fainter under basal conditions [42]. This effect is independent of protein synthesis, and it seems mediated by a microtubule-dependent mechanism of reinsertion of pre-synthesized cotransporters [48]. In contrast, disruption of the microtubule network does not affect in the rapid downregulation of the cotransporter in response to acute administration of high P_i diet. Contrary to the acute response, the upregulation induced by chronic P_i deprivation is associated with an increase in NaPi-IIa mRNA, as compared with levels for the chronic high P_i intake [42]. This effect, at least in OK cells, is rather small and might be due to an increased mRNA stability, rather than to a stimulated transcription rate [49, 50].

Thyroid Hormone

Thyroid hormones strongly influence the growth and differentiation of several tissues. In the adult kidney, triiodothyronine (T_3) stimulates renal P_i reabsorption via upregulation of NaPi-IIa. Such activation has been reported both in rats as well as in OK cells [20, 51]. As mentioned before, a role of T_3 in maturation of NaPi-IIa during weaning has also been suggested [46].

Growth Hormone and Insulin-Like Growth Factor-1

Administration of growth hormone increases the rate of renal P_i reabsorption via upregulation of NaPi-IIa. Most probably this effect is not direct, but instead it seems mediated by insulin-like growth factor-1 [18, 52]. Insulin-like growth factor-1 has been reported to directly stimulate renal P_i transport via both apical and basolateral receptors, although the magnitude of the apical effect was greater than the basolateral [53]. Signalling through the basolateral but not through the apical site results in phosphorylation of the receptor. This finding suggests that whereas both apical and basolateral receptors can upregulate NaPi-IIa, they do not use the same signal transduction mechanism.

Insulin

Insulin has an antiphosphaturic effect in vivo due to stimulation of renal P_i reabsorption [54]. In contrast, an acute decrease in plasma insulin leads to a decrease in renal reabsorption of P_i. In vitro studies on isolated brush border membranes from insulin-treated animals have reported that this results in an increase in the Na^+-dependent P_i cotransport [55]. Based on experiments performed with OK cells, the stimulatory effect of insulin seems to be mediated by changes in the intracellular Ca^{2+} levels [56]. It has been suggested that insulin also plays a role in the rapid renal adaptation to dietary deprivation of P_i [57].

Acid-Base Status

The systemic acid-base status influences renal P_i reabsorption: an alkaline intratubular pH stimulates, while chronic metabolic acidosis decreases the renal P_i reabsorption. Chronic metabolic acidosis in rats induces a decrease in Na^+/P_i cotransport activity, associated with a reduction in type-IIa cotransporter protein and mRNA [58]. This effect is probably due to changes in glucocorticoid levels [59].

Transport Kinetics

Previous studies on P_i transport using tracer flux assays in both perfused renal proximal tubules and the brush border membrane vesicle (BBMV) preparation established important kinetic features that we can now specifically attribute to the NaPi-IIa protein: (i) P_i uptake is Na^+-dependent with cooperative Na^+ activation involving more than one Na^+ ion; (ii) uptake is inhibited by acidification of the external medium, and (iii) this most likely involves preferential transport of divalent P_i.

The advent of expression cloning of renal Na^+/P_i cotransporters enabled the identification of the proteins responsible for P_i uptake, and their overexpression in cell lines and the *X. laevis* oocyte [23], and enabled the NaPi-IIa-specific transport kinetics to be studied with minimal contamination from endogenous Na^+-coupled transport systems. Whereas renal epithelial cell lines transfected with NaPi-IIa isoforms are a convenient means for detailed study of trafficking and regulation because of the polarized morphology and response to PTH (see above), the *X. laevis* oocyte expression system has been used almost exclusively to study transport kinetics.

A significant breakthrough came with the confirmation, as suggested by earlier in vivo studies [60] and BBMV studies [61, 62], that type-IIa-mediated Na^+/P_i cotransport is an electrogenic mechanism, i.e. for each P_i molecule transported, a net charge is transferred across the membrane [63, 73]. This property has important consequences for understanding transport kinetics because, in addition to an inwardly directed Na^+ gradient, membrane potential also provides a driving force for P_i reabsorption.

The high levels of protein expression possible with the oocyte system have allowed transport kinetics to be studied on a single cell under controlled conditions using conventional electrophysiological techniques. Under voltage clamp conditions, in which the membrane potential is held at a predefined value, and the P_i-induced change in transmembrane current is used as a measure of transport activity, several NaPi-IIa isoforms were recently characterized in terms of P_i and Na^+ affinity, pH sensitivity and substrate inhibition [63–66]. The basic findings reported in these studies are in general agreement with those from uptake studies using the oocyte expression system [23] and BBMV isolated from native tissue [67].

Steady-State Activation

At all membrane potentials in the range of –100 to 0 mV, the P_i activation shows a Michaelis-Menten dose response which is consistent with one P_i-binding site. The apparent affinity for P_i (K_m^{Pi}) ≈ 0.1 m*M* decreases with membrane depolarization as well as reduced external Na^+. This behavior is consistent with an ordered binding scheme. External Na^+ also determines the apparent maximum transport rate (V_{max}) of P_i activation, which is consistent with Na^+ being the last substrate to bind. In contrast, the Na^+-activation dose response is sigmoidal (apparent K_m^{Na} ≈ 50 m*M*), with a Hill coefficient in the range of 2–3, which indicates that more than one Na^+ ion binds in a cooperative manner. Although K_m^{Na} is dependent on P_i, the predicted apparent V_{max} at saturating Na^+ is not. These findings emphasize the interactive role that extrinsic variables such as membrane potential and substrate concentration play in determining phenomenological parameters like V_{max} and K_m. These steady-state kinetic parameters have been found to be similar among several renal NaPi-IIa isoforms [63–66, 68]. Recently, using simultaneous uptake and electrophysiology, the preferred P_i species cotransported was shown to be divalent P_i and the Na^+:P_i stoichiometry was confirmed as 3:1 [73]. These findings are consistent with a transfer of one net charge per transport cycle.

Slippage

Another functional property is 'slippage', whereby NaPi-IIa behaves as a uniport for Na^+ ions, i.e. in the absence of external P_i, when the carrier is loaded with one Na^+ ion that can subsequently translocate. This mode of operation appears to be a common feature of a number of Na^+-driven cotransporters. Although its physiological significance is unclear, since this mode would most likely only contribute to net charge transfer at very low luminal P_i, it is important for our understanding of the transport cycle and offers a useful indicator for identification of a functionally important site within the NaPi-IIa protein (see below).

Inhibition of Type-IIa Na^+/P_i Cotransport

pH. One of the hallmarks of NaPi-IIa cotransport kinetics is the increased transport activity with increased external pH. Some of the reduced transport at acidic pH in vivo might be attributable to a tritration effect of P_i depending on the luminal P_i and pH. However, a major contribution of the inhibition observed with exogenously expressed NaPi-IIa can be attributed to a competitive inhibition of the final Na^+ binding and an interaction of protons with the empty carrier that largely affects the voltage dependence of the cotransporter [69]. These interpretations are in agreement with previous studies performed with BBMVs [67].

Forscarnet. This antiviral agent (phosphonoformic acid, PFA) is a known competitive inhibitor of brush border Na^+/P_i cotransport [70]. Electrophysiological studies [68] have demonstrated that PFA competitively inhibits the NaPi-IIa P_i-induced electrogenic response, but does not substitute for P_i as substrate. Other studies [27, 65] have shown that PFA also blocks the slippage mode, which supports the notion that it interacts with the P_i-binding transition (fig. 3).

Presteady-State Relaxations

In addition to the steady-state kinetic characterization, NaPi-IIa kinetics can be described in terms of so-called presteady-state relaxations. If an oocyte expressing NaPi-IIa protein is subjected to a rapid change in membrane voltage, charge movements can be resolved which relax with time constants in the 4- to 15-millisecond range. These charge movements are directly correlated with Na^+/P_i transport activity [65, 66] and result from movement of changes within the transmembrane electric field. Analysis of presteady-state kinetics led to the identification of voltage-sensitive partial reactions. From such analyses of relaxations recorded from oocytes expressing the flounder renal/intestinal type-IIb isoform [66] and the rat renal type-IIa isoform NaPi-IIa [65], we have established that the charge movements that confer voltage sensitivity arise from the empty carrier and first Na^+ binding. In contrast, the P_i- and final Na^+-binding steps are voltage-independent, which suggests that these binding sites are not located within the transmembrane electric field. Presteady-state charge movements can be fitted to a Boltzmann model which allows an estimation of the transport turn-

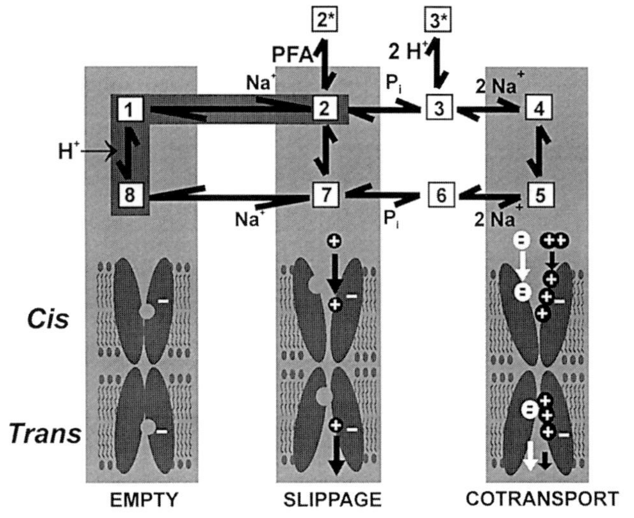

Fig. 3. Kinetic scheme for NaPi-IIa. The states (numbered 1–8) represent unique molecular conformations, only one of which can be occupied at any time, that the protein assumes during the transport cycle. The molecule can either face outward *(cis)* or inward *(trans)*, depending on the availability of substrate and the membrane potential, as indicated in the cartoon representation. Three modes of operation that involve transmembrane reorientation, as revealed by steady-state and presteady-state studies, are indicated by dark shading: empty carrier (8 ⟷ 1); slippage (2 ⟷ 7) and cotransport (4 ⟷ 5). Two voltage dependent transitions have been identified: empty carrier (8 ⟷ 1), first external Na^+ binding/release step (1 ⟷ 2). The empty carrier is assumed to carry an apparent charge of −1. Under physiological conditions, substrate binding on the *cis* side proceeds in an ordered sequence to form, together with the empty carrier, a neutral 'complex' comprising one divalent P_i and three Na^+ ions, which then translocates to the *trans* side. The voltage dependence of the last Na^+-binding/release step on the *trans* side (7 ⟷ 8) has not been characterized and the order of binding/debinding for the other transitions on the *trans* side is not known. PFA binding places system in state 2*, which when occupied, prevents the slippage and cotransport modes. pH-sensitive transitions involve competition with the last Na^+ binding, modelled as a proton binding step (3 ⟷ 3*), and the empty carrier translocation step (1 ⟷ 8).

over. Typically turnover numbers in the range of 10–20 s^{-1} are obtained for NaPi-IIa, which are in agreement with other Na^+-coupled cotransporters [71].

An Ordered Kinetic Scheme for NaPi-IIa

The experimental findings from both tracer and electrophysiological studies have led to the proposal of an alternating access type model for type-IIa Na^+-coupled P_i cotransport as shown in figure 3. Similar schemes have also been proposed for other Na^+-coupled cotransport systems [72], and suggest that common molecular mechanisms are involved, at least in terms of co-substrate activation, despite the lack of protein homology at the structural level. In figure 3, the empty carrier has an apparent valency of −1 that translocates between the *cis* or *trans* face of the membrane under the influence of the membrane potential, thereby making a Na^+-binding site available at the respective side. In reality this apparent valency might be contributed by the concerted movement of many charged residues; however, the presently available measurement techniques do not allow more precise resolution. Binding of substrates occurs in an ordered manner commencing with a single Na^+ ion, which increases the affinity for divalent P_i. In the absence of P_i, the protein operates in the slippage mode. Binding of P_i is followed by a second Na^+-binding step involving two Na^+ ions before translocation of a neutral complex takes place [65].

As determined experimentally, this scheme predicts the net inward transfer of one positive charge per transport cycle under physiological conditions [37]. At present, the model assumes mirror symmetry, and details of the transitions on the *trans* face remain to be determined. Simulations using rate constants, some of which have been determined from measurements, predict all the basic kinetic properties of the NaPi-IIa cotransporter [27, 65, 66, 73], and thereby support the applicability of this scheme in interpretation of data from structure-function studies.

Structure-Function Studies

To identify which structural elements and domains of the NaPi-IIa protein are responsible for specific kinetic properties such as substrate affinity, voltage dependence and pH sensitivity, a combination of biochemical and molecular biological approaches, together with functional assay (flux measurements and electrophysiology) can be used. This is illustrated by two recent examples from our laboratory.

Cysteine Scanning Identifies Functionally Important Residues

The substituted cysteine accessibility method combined with chemical modification of the accessible cysteines can be used to identify functionally important residues in membrane proteins [74]. For example, the mutant NaPi-IIa cotransporter S460C, in which a native serine has been replaced with a cysteine at site 460 in the 3rd extracellular loop (fig. 1) shows essentially normal wild-type kinetics. However, after chemical modification of this cysteine by MTSEA, P_i cotransport, but not slippage, is fully suppressed [27]. Biophysical characterization of this construct indicated that this site is involved in conformational changes in the empty carrier and such findings support the alternating access concept [72]. Moreover, subsequent cysteine mapping in the neighboring region has identified other residues which affect kinetic characteristics such as voltage dependence and pH sensitivity [75].

Chimeras Used to Identify a Site Which Confers pH Sensitivity

Chimeric constructs, made using complementary sections of isoforms of the NaPi-II proteins that have different kinetic characteristics, can be used to identify the regions that confer the respective kinetic property. For example, murine NaPi-IIb, when expressed in *X. laevis* oocytes shows little pH dependence [2] compared with its renal counterpart. Interpretation of pH-dependent P_i uptake data from chimeras, made using complementary regions from the N-terminal and C-terminal halves, led to the location of a pH-sensitive site in the C-terminal half of the protein within the putative linker region between transmembrane domains 4 and 5 [76] (fig. 1). A subsequent sequence comparison of amino acids in this region revealed that three charged amino acids in the IIa protein are represented by neutral amino acids in the IIb protein, and point mutations have confirmed that the former are necessary for conferring pH sensitivity to NaPi-IIa [76].

Acknowledgements

The authors acknowledge financial support from the Swiss National Science Foundation (SNF 31-46523), the Hartmann Müller-Stiftung (Zurich), the Olgar Mayenfisch-Stiftung (Zurich) and the Schweizerischer Bankgesellschaft (Zurich; Bu 704/7-1).

References

1 Murer H, Hernando N, Forster I, Biber J: Proximal tubular phosphate reabsorption: molecular mechanisms. Physiol Rev 2000, in press.
2 Hilfiker H, Hattenhauer O, Traebert M, Forster I, Murer H, Biber J: Characterization of a murine type II sodium-phosphate cotransporter expressed in mammalian small intestine. Proc Natl Acad Sci USA 1998;95:14564–14569.
3 Custer M, Lötscher M, Biber J, Murer H, Kaissling B: Expression of Na-Pi cotransport in rat kidney: Localization by RT-PCR and immunohistochemistry. Am J Physiol 1994;35:F767–F774.
4 Traebert M, Hattenhauer O, Murer H, Kaissling B, Biber J: Expression and localization of a type II sodium-phosphate cotransporter in murine alveolar epithelial cells. Am J Physiol 2000, in press.
5 Kavanaugh MP, Kabat D: Identification and characterization of a widely expressed phosphate transporter/retrovirus receptor family. Kidney Int 1996;49:959–963.
6 Oberbauer R, Schreiner GF, Biber J, Murer H, Meyer TW: In vivo suppression of the renal Na^+/Pi cotransporter by antisense oligonucleotides. Proc Natl Acad Sci USA 1996;93:4903–4906.
7 Werner A, Kempson SA, Biber J, Murer H: Increase of Na/Pi-cotransport encoding mRNA in response to low Pi diet in rat kidney cortex. J Biol Chem 1994;269:6637–6639.
8 Beck L, Karaplis AC, Amizuka N, Hewson AS, Ozawa H, Tenenhouse HS: Targeted inactivation of Npt2 in mice leads to severe renal phosphate wasting, hypercalciuria, and skeletal abnormalities. Proc Natl Acad Sci USA 1998;95:5372–5377.
9 Murer H, Biber J: Molecular mechanisms of renal apical Na/phosphate cotransport. Annu Rev Physiol 1996;58:607–618.
10 Murer H, Biber J: Renal tubular phosphate: Cellular mechanism; in Seldin DW, Giebisch G (eds): The Kidney, Physiology and Pathophysiology. New York, Raven Press, 1992, pp 2481–2509.
11 Murer H, Biber J: A molecular view of proximal tubular inorganic phosphate (Pi) reabsorption and its regulation. Pflügers Arch 1997;433:279–389.
12 Murer H, Forster I, Hilfiker H, Pfister M, Kaissling B, Lötscher M, Biber J: Cellular/molecular control of renal Na/Pi-cotransport. Kidney Int Suppl 1998;53:S2–S10.
13 Tenenhouse HS: Recent advances in epithelial sodium-coupled phosphate transport. Curr Opin Nephrol Hypertens 1999;8:407–414.
14 Hartmann CM, Hewson AS, Kos CH, Hilfiker H, Soumounou Y, Murer H, Tenenhouse HS: Structure of murine and human renal type II Na-phosphate cotransporter genes (Npt2 and NPT2). Proc Natl Acad Sci USA 1996;93:7409–7414.
15 Taketani Y, Miyamoto K, Tanaka K, Takai K, Chikamori M, Tatsumi S, Segawa H, Yamamoto H, Morita K, Takeda E: Gene structure and functional analysis of the human Na^+/phosphate co-transporter. Biochem J 1997;324:927–934.
16 Hilfiker H, Hartmann CM, Stange G, Murer H: Characterization of the 5′-flanking region of the OK cell type II Na-Pi cotransporter gene. Am J Physiol 1998;274:F197–F204.
17 Hilfiker H, Kvieticova I, Hartmann CM, Stange G, Murer H: Characterization of the human type II Na/Pi-cotransporter promoter. Pflügers Arch 1998;436:591–598.
18 Jehle AW, Forgo J, Biber J, Lederer E, Krapf R, Murer H: IGF-I and vanadate stimulate Na/Pi-cotransport in OK cells by increasing type II Na/Pi-cotransporter protein stability. Pflügers Arch 1998;437:149–154.
19 Taketani Y, Segawa H, Chikamori M, Morita K, Tanaka K, Kido S, Yamamoto H, Iemori Y, Tatsumi S, Tsugawa N, Okano T, Kobayashi T, Miyamoto K, Takeda E: Regulation of type II renal Na^+-dependent inorganic phosphate transporters by 1,25-dihydroxyvitamin D_3. Identification of a vitamin D-responsive element in the human NAPI-3 gene. J Biol Chem 1998;273:14575–14581.
20 Alcalde AI, Sarasa M, Raldua D, Aramayona J, Morales R, Biber J, Murer H, Levi M, Sorribas V: Role of thyroid hormone in regulation of renal phosphate transport in young and aged rats. Endocrinology 1999;140:1544–1551.
21 Ritthaler T, Traebert M, Lötscher M, Biber J, Murer H, Kaissling B: Effects of phosphate intake on distribution of type II Na/P_i cotransporter mRNA in rat kidney. Kidney Int 1999;55:976–983.

22 Tatsumi S, Miyamoto K, Kouda T, Motonaga K, Katai K, Ohkido I, Morita K, Segawa H, Tani Y, Yamamoto H, Taketani Y, Takeda E: Identification of three isoforms for the Na^+-dependent phosphate cotransporter (NaPi-2) in rat kidney. J Biol Chem 1998;273:28568–28575.
23 Magagnin S, Werner A, Markovich D, Sorribas V, Stange G, Biber J, Murer H: Expression cloning of human and rat renal cortex Na/Pi cotransport. Proc Natl Acad Sci USA 1993;90: 5979–5983.
24 Biber J, Custer M, Magagnin S, Hayes G, Werner A, Lötscher M, Kaissling B, Murer H: Renal Na/Pi-cotransporters. Kidney Int 1995; 49:981–985.
25 Xiao Y, Boyer CJC, Vincent E, Dugrè A, Vachon V, Potier M, Bèliveau R: Involvement of disulphide bonds in the renal sodium/phosphate cotransporter NaPi-2. Biochem J 1997; 323:401–408.
26 Lambert G, Traebert M, Hernando N, Biber J, Murer H: Studies on the topology of the renal type II NaPi-cotransporter. Pflügers Arch 1999;437:972–978.
27 Lambert G, Forster IC, Stange G, Biber J, Murer H: Properties of the mutant Ser-460-Cys implicate this site in a functionally important region of the type IIa Na^+/P_i cotransporter protein. J Gen Physiol 1999;114:637–651.
28 Hayes G, Busch A, Lötscher M, Waldegger S, Lang F, Verrey F, Biber J, Murer H: Role of N-linked glycosylation in rat renal Na/Pi-cotransport. J Biol Chem 1994;269:24143–24149.
29 Hernando N, Sheikh S, Karim-Jimenez Z, Galliker H, Forgo J, Biber J, Murer H: Asymmetrical targeting of type IIa Na/Pi-cotransporters in renal and intestinal epithelial cell lines. Am J Physiol 2000;278:F361–F368.
30 Karim-Jimenez Z, Hernando N, Biber J, Murer H: Molecular signals for the apical expression of type IIa NaPi-cotransporters. In preparation.
31 Kempson SA, Lötscher M, Kaissling B, Biber J, Murer H, Levi M: Parathyroid hormone action on phosphate transporter mRNA and protein in rat renal proximal tubules. Am J Physiol 1995;268:F784–F791.
32 Pfister MF, Lederer E, Forgo J, Ziegler U, Lötscher M, Quabius ES, Biber J Murer H: Parathyroid hormone-dependent degradation of type II Na+/Pi cotransporters. J Biol Chem 1997;272:20125–20130.
33 Pfister MF, Ruff I, Stange G, Ziegler U, Lederer E, Biber J, Murer H: Parathyroid hormone leads to the lysosomal degradation of the renal type II Na/Pi cotransporter. Proc Natl Acad Sci USA 1998;95:1909–1914.
34 Keusch I, Traebert M, Lötscher M, Kaissling B, Murer H, Biber J: Parathyroidhormone and dietary phosphate provoke lysosomal routing of the proximal tubular Na/Pi-cotransporter type II. Kidney Int 1998;54:1224–1232.
35 Lötscher M, Scarpetta Y, Levi M, Halaihel N, Wang H, Zajicek HZ, Biber J, Murer H, Kaissling B: Rapid downregulation of rat renal Na/Pi cotransporter in response to parathyroid hormone involves microtubule rearrangement. J Clin Invest 1999;104:483–494.
36 Hayes G, Busch A, Lang F, Biber J, Murer H: Protein kinase C consensus sites and the regulation of the renal Na/Pi-cotransport (NaPi-2) expressed in *Xenopus laevis* oocytes. Pflügers Arch 1995;430:819–824.
37 Forster I, Traebert M, Jankowski M, Stange G, Biber J, Murer H: Protein kinase C activators induce membrane retrieval ot type II Na^+-phosphate cotransporters expressed in *Xenopus* oocytes. J Physiol (Lond) 1999;517:327–340.
38 Wagner CA, Raber G, Waldegger S, Oddwald H, Biber J, Murer H, Busch AE, Lang F: Regulation of the human brush border Na^+/phosphate cotransporter (NaPi-3) expressed in *Xenopus* oocytes by intracellular calcium and protein kinase C. Cell Physiol Biochem 1996;6: 105–111.
39 Mellman I: Endocytosis and molecular sorting. Annu Rev Cell Dev Biol 1996;12:575–625.
40 Traebert M, Roth J, Biber J, Murer H, Kaissling B: Internalization of proximal tubular type II Na/Pi-cotransporter by parathyroid hormone: An immunogold electron microscopy study. Am J Physiol 2000;278:F148–F154.
41 Hernando N, Forgo J, Biber J, Murer H: PTH-induced downregulation of the type IIa cotransporter is independent of known endocytic motifs. J Am Soc Nephrol, in press.
42 Levi M, Lötscher M, Sorribas V, Custer M, Arar M, Kaissling B, Murer H, Biber J: Cellular mechanisms of acute and chronic adaptation of rat renal Pi transporter to alterations in the dietary Pi. Am J Physiol 1994;267:F900–F908.
43 Pfister MF, Hilfiker H, Forgo J, Lederer E, Biber J, Murer H: Cellular mechanisms involved in the acute adaptation of OK cell Na/Pi-cotransport to high- or low-P_i medium. Pflügers Arch 1998;435:713–719.
44 Levi M, Shayman JA, Abe A, Gross SK, McCluer RH, Biber J, Murer H, Lötscher M, Cronin RE: Dexamethasone modulates rat renal brush border membrane phosphate transporter mRNA and protein abundance and glycosphingolipid composition. J Clin Invest 1995;96:207–216.
45 Traebert M, Lötscher M, Aschwanden R, Ritthaler T, Biber J, Murer H, Kaissling B: Distribution of the sodium/phosphate transporter during postnatal ontogeny of the rat kidney. J Am Soc Nephrol 1999;10:1407–1415.
46 Euzet S, Lelièvre-Pègorier M, Merlet-Bènichou C: Effect of 3,5,3)-triiodothyronine on maturation of rat renal phosphate transport: Kinetic characteristics and phosphate transporter messenger ribonucleic acid and protein abundance. Endocrinology 1996;137:3522–3530.
47 Silverstein DM, Barac-Nieto M, Murer H, Spitzer A: A putative growth-related renal Na^+-Pi cotransporter. Am J Physiol 1997;273: R928–R933.
48 Lötscher M, Kaissling B, Biber J, Murer H, Levi M: Role of microtubules in the rapid regulation of renal phosphate transport in response to acute alterations in dietary phosphate content. J Clin Invest 1997;99:1302–1312.
49 Markovich D, Verri T, Sorribas V, Forgo J, Biber J, Murer H: Regulation of opossum kidney (OK) cell Na/Pi cotransport by Pi deprivation involves mRNA stability. Pflügers Arch 1995;430:459–463.
50 Saxena S, Dansby L, Allon M: Adaptation to phosphate depletion in opossum kidney cells. Biochem Biophys Res Commun 1995;216: 141–147.
51 Sorribas V, Markovich D, Verri T, Biber J, Murer H: Thyroid hormone stimulation of Na/Pi-cotransport in opossum kidney cells. Pflügers Arch 1995;431:266–271.
52 Hammerman MR: The growth hormone-insulin-like growth factor axis in kidney. Am J Physiol 1989;257:F503–F514.
53 Quigley R, Baum M: Effects of growth hormone and insulin-like growth factor I on rabbit proximal convoluted tubule transport. J Clin Invest 1991;88:368–374.
54 DeFronzo RA, Cooke RC, Reubin A, Faloona GR, Davis PJ: Effect of insulin on renal handing of sodium, calcium and phosphate in man. J Clin Invest 1975;55:845–855.
55 Hammermann MR, Rogers S, Hansen VA, Gavin JR: Insulin stimulates P_i transport in brushborder vesicles from proximal tubular segments. Am J Physiol 1984;247:E616–E624.
56 Mohan AI, McAteer JA, Kempson SA: Insulin stimulates phosphate transport in opossum kidney epithelial cells. Am J Physiol 1990;258: F1592–F1598.
57 Mohan AI, Woods RE, Breedlove DK, Kempson SA: Renal adaptation to low-phosphate diet in diabetic rats. Am J Physiol 1992;262: F731–F736.
58 Ambuhl PM, Zajicek HK, Wang H, Puttaparthi K, Levi M: Regulation of renal phosphate transport by acute and chronic metabolic acidosis in rats. Kidney Int 1998;53:1288–1298.
59 Boross M, Kinsella J, Cheng L, Sacktor B: Glucocorticoids and metabolic acidosis-induced renal transports of inorganic phosphate, calcium and NH4. Am J Physiol 1986;250:F827–F833.
60 Samarzija I, Molnar V, Frömter E: pH-dependence of phosphate in rat renal proximal tubule. Proc Eur Dial Transpl Assoc 1982;19: 779–783.
61 Burckhardt G, Stern H, Murer H: The influence of pH on phosphate transport into rat renal brush border membrane vesicles. Pflügers Arch 1981;390:191–197.
62 Béliveau R, Strévey J: Phosphate transport in kidneys: Effect of transmembrane electrical potential. Am J Physiol 1991;261:F663–F669.
63 Busch AE, Waldegger S, Herzer T, Biber J, Markovich D, Hayes G, Murer H, Lang F: Electrophysiological analysis of Na^+/P_i cotransport mediated by a transporter cloned from rat kidney in *Xenopus* oocytes. Proc Natl Acad Sci USA 1994;91:8205–8208.

64 Hartmann CM, Wagner CA, Busch AE, Markovich D, Biber J, Lang F, Murer H: Transport characteristics of a murine renal Na/Pi-cotransporter. Pflügers Arch 1995;430:830–836.
65 Forster I, Hernando N, Biber J, Murer H: The voltage dependence of a cloned mammalian renal type II Na^+/P_i cotransporter (NaPi-2). J Gen Physiol 1998;112:1–18.
66 Forster IC, Wagner CA, Busch AE, Lang F, Biber J, Hernando N, Murer H, Werner A: Electrophysiological characterization of the flounder type II Na^+/P_i cotransporter (NaPi-5) expressed in *Xenopus laevis* oocytes. J Membr Biol 1997;160:9–25.
67 Amstutz M, Mohrmann M, Gmaj P, Murer H: Effect of pH on phosphate transport in rat renal brush border vesicles. Am J Physiol 1985;248: F705–F710.
68 Busch AE, Wagner CA, Schuster A, Waldegger S, Biber J, Murer H, Lang F: Properties of electrogenic P_i transport by NaPi-3, a human renal brush border Na^+/P_i transporter. J Am Soc Nephrol 1995;6:1547–1551.
69 Forster I, Biber J, Murer H: Proton-sensitive transitions of renal type II Na^+-coupled phosphate cotransporter kinetics. Biophys J 2000; 79: in press.
70 Kempson SA: Novel specific inhibitors of epithelial phosphate transport. News Physiol Sci 1988;3:154–157.
71 Loo DF, Hazama A, Supplisson S, Turk E, Wright EM: Relaxation kinetics of the Na^+/glucose cotransporter. Proc Natl Acad Sci USA 1993;90:5767–5771.
72 Loo DDF, Hirayama BA, Gallardo EM, Lam JT, Turk E, Wright EM: Conformational changes couple Na^+ and glucose transport. Proc Natl Acad Sci USA 1998;95:7789–7794.
73 Forster I, Loo DDF, Eskandari S: Stoichiometry and Na^+ binding cooperativity of rat and flounder renal type II Na^+-P_i cotransporters. Am J Physiol 1999;276:F644–F649.
74 Akabas MH, Stauffer DA, Xu M, Karlin A: Acetylcholine receptor channel structure probed in cysteine-substitution mutants. Science 1992;258:307–310.
75 Lambert G, Forster I, Stange G, Biber J, Murer H: Cysteine scanning reveals novel structure-function features in the predicted 3rd extracellular loop of the type IIa Na/P_i-CT. In preparation.
76 de la Horra C, Hernando N, Lambert G, Forster I, Biber J, Murer H: Molecular determinants of pH sensitivity of the type IIa Na/P_i-cotransporter. J Biol Chem 2000;275:6284–6287.
77 Chen X-Z, Shayakul C, Berger UV, Tian W, Hediger M: Characterization of a rat Na+-dicarboxylate cotransporter. J Biol Chem 1998; 273:20972–20981.
78 Eskandari S, Loo DDF, Dai G, Levy O, Wright EM, Carrasco N: Thyroid Na^+/I^- symporter. Mechanism, stoichiometry, and specificity. J Biol Chem 1997;272:27320–27328.
79 Kessler M, Semenza G: The small-intestinal Na^+, *D*-glucose contransporter: An asymmetric gated channel (or pore) responsive to delta psi. J Membr Biol 1983;76:27–56.
80 Parent L, Supplisson S, Loo DDF, Wright EM: Electrogenic properties of the cloned Na^+/glucose cotransporter. II. A transport model under nonrapid equilibrium conditions. J Membr Biol 1992;125:63–79.
81 Seifert SA, Hsiao SC, Murer H, Biber J, Kempson SA: Renal endosomal phosphate (Pi) transport in normal and diabetic rats and response to chronic Pi deprivation. Cell Biochem Funct 1997;15:9–14.

Experimental Nephrology

Author Index Vol. 8, No. 6, 2000

Biber, J. 366
Deen, P.M.T. 326
Forster, I.C. 366
Friedman, P.A. 343
Hernando, N. 366
Kamsteeg, E.J. 326
Kunzelmann, K. 332
Marumo, F. 361
Murer, H. 366
Nitschke, R. 332
Oh, Y.S. 320
Os, C.H. van 326
Thakker, R.V. 319, 351
Uchida, S. 361
Warnock, D.G. 320

Subject Index Vol. 8, No. 6, 2000

Autosomal dominant forms 326
– recessive forms 326
Calcium channels 343
– excretion 343
CFTR processing 332
Chloride channel gene 351
Countercurrent system 361
Cystic fibrosis 332
– – transmembrane conductance regulator 332
Diuretics 343
Epithelial Na^+ channel 320
Gene targeting 361
Genetic defects 320
Gitelman's syndrome 343
Henle's loop, thin limb 361
Hypercalciuria 351
Hypertension 320
Ion channels 343
Kidney stones 351
Low molecular weight proteinuria 351
Nephrogenic diabetes insipidus 326
– – –, mice 361
Nephron segments 343
Oligomerization 326
Parathyroid hormone 343, 366
Pathophysiology 320
Pi transport 366
Protein trafficking 366
Proximal tubule 366
Traffic 332
Trafficking 326
Water channels 326

KARGER

Fax + 41 61 306 12 34
E-Mail karger@karger.ch
www.karger.com

Accessible online at:
www.karger.com/journals/exn

Experimental
Nephrology

Author Index Vol. 8, 2000

Aguilar, M. 161
Akamizu, T. 235
Al-Awqati, Q. 66
Álvarez, J. 299
Anders, H.-J. 181
Arima, S. 304
Arrebola, F. 161
Auwardt, R.B. 144, 291

Badr, K.F. 14, 258
Bailey, M.A. 44
Beaufils, H. 104
Biber, J. 366
Birck, R. 173
Bitz, H. 266, 275
Boutros, M. 123
Braun, C. 173, 283
Brezis, M. 266, 275
Bruijn, J.A. 77
Burrow, C.R. 219
Burtin, M. 104

Carbajo, E. 299
Cattell, V. 128
Ceyssens, B. 91
Chen, C.-G. 144, 291
Chevalier, R.L. 115
Clements, R.S. 135
Cobo, A. 299
Cohen, M.P. 135
Coletta, I. 37
Conzelmann, T. 283

Daha, M.R. 203
Daniel, C. 52
Darmon, D. 266, 275
De Bleser, P. 91
De Craemer, D. 91
Deen, P.M.T. 326
De Lucio-Cazaña, F.J. 97
Doi, S.Q. 235

Fernández, M. 299
Fernández, P. 299
Forster, I.C. 366
Frey, B. 52
Friedlander, G. 104
Friedman, P.A. 343
Fukuoka, K. 84

García-Chicano, M.J. 161
Geerts, A. 91

Goldfarb, M. 266, 275
Gómez-Morales, M. 161
Goodyer, P. 123
Guglielmotti, A. 37

Han, D.C. 135
Hayashi, M. 215
Heer, E. de 77
Heikkilä, P. 1
Hekmati, M. 104
Hellemans, K. 91
Hernando, N. 366
Heyman, S.N. 266, 275
Hikita, C. 66
Holt, S. 72
Hud, E. 135

Iglesias-de la Cruz, M.C. 97
Imai, M. 252
Imai, Y. 28
Ishibashi, K. 252
Ishizuka, S. 84
Ito, S. 28, 304

Jouanneau, C. 104

Kaizuma, S. 28
Kamiya, Y. 84
Kamsteeg, E.J. 326
Kanao, K. 20
Kariyil, J.T. 235
Kashihara, N. 20
Kim, G.H. 235
Kirchengast, M. 283
Kittelberger, A.M. 152
Knoll, T. 173
Kooten, C. van 203
Kopchik, J.J. 235
Koshiyama, H. 235
Krajewski, S. 115
Kreft, B. 244
Krzossok, S. 173
Kunzelmann, K. 332

Lelièvre-Pégorier, M. 63
Luca, M.E. 77

McRae, J.L. 291
Maeshima, Y. 20
Makino, H. 20
Mancini, F. 37
Mantovani, A. 37

Marti, H.P. 52
Marumo, F. 361
Masseroli, M. 161
Masson, N. 135
Matsubara, M. 28
Medina, A. 299
Merlet-Bénichou, C. 63
Michael, A.E. 44
Michimata, M. 28
Milanese, C. 37
Miyake, N. 84
Monkawa, T. 215
Montero, A. 14, 258
Moore, K. 72
Moral, R.G. del 161
Mudge, S.J. 144, 291
Munger, K.A. 258
Murakami, T. 20
Murer, H. 366
Murphy, B.F. 291

Nagasawa, T. 84
Nassar, G.M. 258
Nitschke, R. 332
Nogae, S. 28

Oh, Y.S. 320
Ohta, M. 28
Oite, T. 312
Okamoto, K. 20
Olmo, A. 161
Ong, A.C.M. 208
Os, C.H. van 326
O'Valle, F. 161

Pagel, H. 244
Paul, L.C. 77
Pfister, M. 52
Phelps, R.G. 226
Pinza, M. 37
Polentarutti, N. 37
Power, D.A. 144, 291

Ramírez, C. 161
Reed, J.C. 115
Reguero, M.E. 161
Revelles, F. 161
Roberton, C.A. 194
Rodríguez, J. 299
Rodríguez-Puyol, D. 97
Rodríguez-Puyol, M. 97
Rohmeiss, P. 173, 283

KARGER
Fax +41 61 306 12 34
E-Mail karger@karger.ch
www.karger.com

Accessible online at:
www.karger.com/journals/exn

Rosen, S. 266, 275
Rozen, R. 123
Ruiz-Torres, M.P. 97

Santos, F. 299
Saruta, T. 215
Sasaki, S. 252
Schaub, M. 283
Schlöndorff, D. 181
Schwartz, G.J. 152
Segel, G.B. 152
Sekikawa, T. 20
Sellitti, D.F. 235
Shah, S.V. 8
Shina, A. 266, 275
Smith, C.D. 115
Soejima, A. 84
Soldo, L. 37
Sugiyama, H. 20
Suzuki, M. 84

Takito, J. 66
Terzi, F. 104
Thakker, R.V. 319, 351
Thompson, A. 44
Thorner, P. 1
Tryggvason, K. 1

Uchida, S. 361
Uda, S. 258
Ueda, N. 8
Unwin, R.J. 44

Van den Branden, C. 91
Verbeelen, D. 91
Vetter, S. 283
Vijayakumar, S. 66
Vyse, T.J. 194

Waddington, S.N. 128
Wal, A.M. van der 77
Warnock, D.G. 320
Wiederholt, T. 66
Wolstenholme, J. 115
Woltman, A.M. 203
Woude, F.J. van der 173, 283

Yamasaki, Y. 20
Yan, L. 66

Ziswiler, R. 52
Ziyadeh, F. 135

Subject Index Vol. 8, 2000

Experimental Nephrology

Acid-base balance 44
Acute renal failure 8, 84, 244, 283, 299
– tubular necrosis 275
Afferent arteriole 304
Alport syndrome 1
Amino acid transport 123
Angiotensin II 161
Animal models 1
Antigen presenting cells 203
Antioxidant enzymes 91
Antisense oligonucleotides 144
Anti-Thy1.1 nephritis 52
AP-1 104
Apoptosis 312
Arachidonic acid 14
Arginase 128
ATP-binding sites 152
Autoantibodies 194
Autoimmune diseases 77
Autoimmunity 194, 226
Autosomal dominant forms 326
– – polycystic kidney disease 208
– recessive forms 326

Backcrossing 181
Band 3 66
Birth weight 63
Bosentan 266, 275
Brush border membrane 252
DL-Buthionine sulfoximine 84

Calcium channels 343
– excretion 343
Caldesmon 20
Caspases 8
Catalase 91
Cct 1 152
CD40 203
Cell death 8
– markers 312
– proliferation 104
Ceramide 8
CFTR processing 332
Chaperonin 152
Chaperonin-containing TCP1 152
Chemokines 37, 203
Chloride channel gene 351
Collagen I 97
– IV 97
Complement 291
Confocal laser scan microscopy 312
Contractile protein 20
Contraction 97
Countercurrent system 361
c-ret 219
CRP-ductin 66
Crush syndrome 72
Cyclosporin A 161
Cystic fibrosis 332
– – transmembrane conductance regulator 332
– kidney disease 115
Cystinuria 123
Cytokines 258
Cytoskeletal protein 20

db/db mouse 135
Diabetes 63, 135
Diuretics 343
DMBT1 66
Double staining 312
Drugs 63

Ebnerin 66
Efferent arteriole 304
Endonucleases 8
Endothelin 266, 283
– 1 161
– 3 161
– A receptor antagonist 283
Endotoxin 266, 275
Epithelial Na^+ channel 320
– polarity 66
Escherichia coli 244
Extracellular matrix 97

Gene expression 91
– targeting 361
– therapy 1
Genetic defects 320
– linkage 194
Gitelman's syndrome 343
Glomerular antigens 226
– injury 20
Glomerulonephritis 14, 52, 128, 181, 226
Glomerulosclerosis 91, 181
Glomerulus 304
Glucocorticoids 44
Glutathione 84
– peroxidase 91
Glycated albumin 135
H^+ ATPase 66
Heme 72
Hemodynamics 266
Henle's loop, thin limb 361
Hensin 66
Heymann nephritis 77
Hydrogen peroxide 97
11β-Hydroxysteroid dehydrogenase 44
Hypercalciuria 351
Hypertension 320
Hypoxia 275

IκB-α 144
Immunocytochemistry 115
Immunohistochemistry 208, 215
Immunostaining 312
Indomethacin 266, 275
Inflammation 128, 203
γ-Interferon 258
Inositol trisphosphate receptor 215
Intercalated cell 66
Interleukin(s) 258
Interleukin-6 37
– receptor 37
Interstitial nephritis 181
Ion channels 343
Iron 72
Isoforms 215
Isolated glomeruli 312
– perfused kidney 275
– – rat kidney 244

Kidney 1, 44, 152, 215, 219
– failure, acute 266, 275
– organogenesis 63
– stones 123, 351
– transplantation 283

L-NAME 266, 275
Laser Doppler 266
– – flowmetry 173
Leukotrienes 258
Lipid peroxidation 72, 84
Lipoxins 14
Low molecular weight proteinuria 351
LTA_4 hydrolase 258

Macrophage 128
Major histocompatibility complex 194
Manoalide 84
MCP-1 37, 144
Medullary thick ascending limb 275

KARGER
Fax + 41 61 306 12 34
E-Mail karger@karger.ch
www.karger.com

Accessible online at:
www.karger.com/journals/exn

Mesangial cell(s) 20, 37, 97, 128, 144, 235, 291
– proliferative glomerulonephritis 52
Metanephros 219
Microcirculation 266
Microdissection 304
Microelectrode 266
Microperfusion 304
Mitogenesis 291
Mouse 194
Mycophenolate mofetil 77

Nephrogenic diabetes insipidus 326
– – –, mice 361
Nephrolithiasis 123
Nephron 219
– number 63
– segments 343
Nephropathy 135
Nephrotoxicity 161
NF-κB 144, 291
Nitric oxide 128, 266
Nonsteroidal anti-inflammatory drugs 173

Oligomerization 326
Optical sectioning 312
Osmolality 28

Parathyroid hormone 343, 366
Pathophysiology 320
Phospholipase A_2 84
Phylogenetic analysis 152
Pi transport 366
PKD1 208
Polycystin 208
Potassium 44
Programmed cell death 115
Proliferation 97
Protein trafficking 366
Proto-oncogenes 104
Proximal straight tubule 252
– tubule 366
Puromycin amino nucleoside nephrosis 52

Rapamycin 52
Rat(s) 283, 299
rBAT 123
Reactive oxygen metabolites 8
– – species 97
Receptor localization 215
Recovery 299
Redox cycling 72
Remnant kidney 91, 104
Renal development 219
– disease 181
– lesions 104
– maturation 115
– medulla perfusion 173
– tubular epithelial cells 8
– water reabsorption 28
Renovascular hypertension 181
Retinoic acid receptor 219
Retinoids 219

SDZ RAD 52
Sodium 44
– restriction 104
Sublytic complement injury 291
Superoxide dismutase 91
Systemic lupus erythematosus 194

Terminal differentiation, plasticity 66
Thyroglobulin 235
Thyroid 235
Thyrotropin receptor 235
Traffic 332
Trafficking 326
Transgenic mice 181
Tubular epithelial cell(s) 115, 203
TUNEL method 312
Type-IV collagen 1

Ultrasonic transit-time flowmetry 173
Urea transport 252
Ureteral obstruction 115
Ureteric bud 219

Vasopressin 28
Vitamin A 63
– – deficiency syndrome 219

Water channel(s) 28, 252, 326
Western analysis 208

Vol. 8, 2000

Experimental Nephrology

Journal of Renal Cell Biology

KARGER

Basel · Freiburg · Paris · London · New York · New Delhi · Bangkok · Singapore · Tokyo · Sydney

S. Karger
Medical and Scientific Publishers
Basel · Freiburg · Paris · London
New York · New Delhi · Bangkok
Singapore · Tokyo · Sydney

Drug Dosage

The authors and the publisher have exerted every effort to ensure that drug selection and dosage set forth in this text are in accord with current recommendations and practice at the time of publication. However, in view of ongoing research, changes in government regulations, and the constant flow of information relating to drug therapy and drug reactions, the reader is urged to check the package insert for each drug for any change in indications and dosage and for added warnings and precautions. This is particularly important when the recommended agent is a new and/or infrequently employed drug.

P.O. Box, CH–4009 Basel (Switzerland)
Printed in Switzerland on acid-free paper by
Reinhardt Druck, Basel

KARGER

Fax + 41 61 306 12 34
E-Mail karger@karger.ch
www.karger.com

Contents Vol. 8, 2000

Experimental Nephrology

No. 1

Minireviews

1 **Animal Models of Alport Syndrome: Advancing the Prospects for Effective Human Gene Therapy**
Heikkilä, P.; Tryggvason, K. (Stockholm); Thorner, P. (Toronto)

8 **Role of Endonucleases in Renal Tubular Epithelial Cell Injury**
Ueda, N.; Shah, S.V. (Little Rock, Ark.)

14 **15-Lipoxygenase in Glomerular Inflammation**
Montero, A.; Badr, K.F. (Atlanta, Ga.)

Original Papers

20 **Caldesmon Isoform Associated with Phenotypic Modulation of Mesangial Cells**
Okamoto, K.; Kashihara, N.; Yamasaki, Y.; Kanao, K.; Maeshima, Y.; Sekikawa, T.; Sugiyama, H.; Murakami, T.; Makino, H. (Okayama)

28 **Topographic Distribution of Aquaporin 2 mRNA in the Kidney of Dehydrated Rats**
Michimata, M.; Nogae, S.; Ohta, M.; Kaizuma, S.; Imai, Y.; Ito, S.; Matsubara, M. (Sendai)

37 **Selective Induction of MCP-1 in Human Mesangial Cells by the IL-6/sIL-6R Complex**
Coletta, I.; Soldo, L. (Roma); Polentarutti, N. (Milano); Mancini, F.; Guglielmotti, A.; Pinza, M. (Roma); Mantovani, A. (Milano); Milanese, C. (Rome)

44 **Effects of Changes in Dietary Intake of Sodium and Potassium and of Metabolic Acidosis on 11β-Hydroxysteroid Dehydrogenase Activities in Rat Kidney**
Thompson, A.; Bailey, M.A.; Michael, A.E.; Unwin, R.J. (London)

52 **Proinflammatory Effects in Experimental Mesangial Proliferative Glomerulonephritis of the Immunosuppressive Agent SDZ RAD, a Rapamycin Derivative**
Daniel, C.; Ziswiler, R.; Frey, B.; Pfister, M.; Marti, H.-P. (Bern)

No. 2

Minireviews

63 **The Number of Nephrons in the Mammalian Kidney: Environmental Influences Play a Determining Role**
Lelièvre-Pégorier, M.; Merlet-Bénichou, C. (Paris)

66 **Phenotypic Plasticity and Terminal Differentiation of the Intercalated Cell: The Hensin Pathway**
Al-Awqati, Q.; Vijayakumar, S.; Takito, J.; Hikita, C.; Yan, L.; Wiederholt, T. (New York, N.Y.)

72 **Pathogenesis of Renal Failure in Rhabdomyolysis: The Role of Myoglobin**
Holt, S.; Moore, K. (London)

Original Papers

77 **Treatment with Mycophenolate Mofetil Attenuates the Development of Heymann Nephritis**
Luca, M.E. (Leiden); Paul, L.C. (Toronto); van der Wal, A.M.; Bruijn, J.A.; de Heer, E. (Leiden)

84 **Simultaneous Inhibition of Renal Phospholipase A_2 and Glutathione Synthesis by Manoalide and *DL*-Buthionine Sulfoximine Induces Acute Tubular Dysfunction in Rats**
Soejima, A.; Ishizuka, S.; Miyake, N.; Fukuoka, K.; Suzuki, M.; Kamiya, Y.; Nagasawa, T. (Tokyo)

91 **Antioxidant Enzyme Gene Expression in Rats with Remnant Kidney Induced Chronic Renal Failure**
Van den Branden, C.; Ceyssens, B.; De Craemer, D.; De Bleser, P.; Hellemans, K.; Geerts, A.; Verbeelen, D. (Brussel)

97 **Phenotypic Modifications of Human Mesangial Cells by Extracellular Matrix: The Importance of Matrix in the Contractile Response to Reactive Oxygen Species**
Iglesias-de la Cruz, M.C.; Ruiz-Torres, M.P.; De Lucio-Cazaña, F.J.; Rodríguez-Puyol, M.; Rodríguez-Puyol, D. (Alcalá de Henares)

104 **Sodium Restriction Decreases AP-1 Activation after Nephron Reduction in the Rat: Role in the Progression of Renal Lesions**
Terzi, F.; Burtin, M.; Hekmati, M.; Jouanneau, C.; Beaufils, H.; Friedlander, G. (Paris)

115 **Chronic Ureteral Obstruction in the Rat Suppresses Renal Tubular Bcl-2 and Stimulates Apoptosis**
Chevalier, R.L.; Smith, C.D.; Wolstenholme, J. (Charlottesville, Va.); Krajewski, S.; Reed, J.C. (La Jolla, Calif.)

90 **Announcement**

No. 3

Minireviews

123 **The Molecular Basis of Cystinuria: An Update**
Goodyer, P.; Boutros, M.; Rozen, R. (Montreal)

128 **Arginase in Glomerulonephritis**
Waddington, S.N.; Cattell, V. (London)

Original Papers

135 **Inhibiting Albumin Glycation Ameliorates Diabetic Nephropathy in the *db/db* Mouse**
Cohen, M.P.; Masson, N.; Hud, E.; Ziyadeh, F.; Han, D.C.; Clements, R.S. (Philadelphia, Pa.)

144 **Inhibition with Antisense Oligonucleotide Suggests that IκB-α Does Not Form a Negative Autoregulatory Loop for NF-κB in Mesangial Cells**
Auwardt, R.B.; Mudge, S.J.; Chen, C.-G.; Power, D.A. (Melbourne)

152 **Cloning of Rabbit Cct6 and the Distribution of the Cct Complex in Mammalian Tissues**
Schwartz, G.J.; Kittelberger, A.M.; Segel, G.B. (Rochester, N.Y.)

161 **Role of Intrarenal Endothelin 1, Endothelin 3, and Angiotensin II Expression in Chronic Cyclosporin A Nephrotoxicity in Rats**
Ramírez, C.; Olmo, A.; O'Valle, F.; Masseroli, M.; Aguilar, M.; Gómez-Morales, M.; Revelles, F.; García-Chicano, M.J.; Arrebola, F.; Reguero, M.E.; del Moral, R.G. (Granada)

173 **Preferential COX-2 Inhibitor, Meloxicam, Compromises Renal Perfusion in Euvolemic and Hypovolemic Rats**
Birck, R.; Krzossok, S.; Knoll, T.; Braun, C.; van der Woude, F.J.; Rohmeiss, P. (Mannheim)

180 **Announcement**

KARGER

Fax + 41 61 306 12 34
E-Mail karger@karger.ch
www.karger.com

Access to full text and tables of contents, including tentative ones for forthcoming issues:
www.karger.com/journals/exn/exn_bk.htm

No. 4–5

Editorial Review

181 **Murine Models of Renal Disease: Possibilities and Problems in Studies Using Mutant Mice**
Anders, H.-J.; Schlöndorff, D. (Munich)

Minireviews

194 **The Genetics of Systemic Lupus erythematosus**
Roberton, C.A.; Vyse, T.J. (London)

203 **Immunological Function of Tubular Epithelial Cells: The Functional Implications of CD40 Expression**
van Kooten, C.; Woltman, A.M.; Daha, M.R. (Leiden)

208 **Polycystin Expression in the Kidney and Other Tissues: Complexity, Consensus and Controversy**
Ong, A.C.M. (Headington)

215 **Cell-Specific Expression of the IP_3 Receptor Gene Family in the Kidney**
Hayashi, M.; Monkawa, T.; Saruta, T. (Tokyo)

219 **Retinoids and Renal Development**
Burrow, C.R. (New York, N.Y.)

226 **Immune Recognition of Glomerular Antigens**
Phelps, R.G. (Edinburgh)

Original Papers

235 **Renal Expression of Two 'Thyroid-Specific' Genes: Thyrotropin Receptor and Thyroglobulin**
Sellitti, D.F. (Bethesda, Md.); Akamizu, T. (Kyoto,); Doi, S.Q.; Kim, G.H. (Bethesda, Md.); Kariyil, J.T. (Washington, D.C.); Kopchik, J.J. (Athens, Ohio); Koshiyama, H. (Amagasaki)

244 **Virulence Factors of *Escherichia coli* Contribute to Acute Renal Failure**
Kreft, B.; Pagel, H. (Lübeck)

252 **Cellular Localization of Aquaporin 7 in the Rat Kidney**
Ishibashi, K.; Imai, M. (Minamikawachi); Sasaki, S. (Tokyo)

258 **Reciprocal Regulation of LTA_4 Hydrolase Expression in Human Monocytes by γ-Interferon and Interleukins 4 and 13: Potential Relevance to Leukotriene Regulation in Glomerular Disease**
Montero, A.; Nassar, G.M.; Uda, S.; Munger, K.A.; Badr, K.F. (Atlanta, Ga.)

266 **Endotoxin-Induced Renal Failure.** I. A Role for Altered Renal Microcirculation
Heyman, S.N.; Darmon, D.; Goldfarb, M.; Bitz, H.; Shina, A. (Jerusalem); Rosen, S. (Boston, Mass.); Brezis, M. (Jerusalem)

275 **Endotoxin-Induced Renal Failure.** II. A Role for Tubular Hypoxic Damage
Heyman, S.N. (Jerusalem); Rosen, S. (Boston, Mass.); Darmon, D.; Goldfarb, M.; Bitz, H.; Shina, A.; Brezis, M. (Jerusalem)

283 **Improved Recovery following Posttransplant Acute Renal Failure in Rat Renal Isografts with an Oral Endothelin-A Receptor Antagonist**
Braun, C.; Vetter, S.; Conzelmann, T.; Schaub, M. (Mannheim); Kirchengast, M. (Ludwigshafen); van der Woude, F.J.; Rohmeiss, P. (Mannheim)

291 **Sublytic Complement Injury Does Not Activate NF-κB, or Induce Mitogenesis in Rat Mesangial Cells**
Mudge, S.J.; McRae, J.L.; Auwardt, R.B.; Murphy, B.F.; Chen, C.-G.; Power, D.A. (Melbourne)

Brief Communication

299 **Prepubertal Rats Are More Resistant to Ischemic Renal Injury and Recover More Rapidly than Adult Rats**
Fernández, M.; Medina, A.; Carbajo, E.; Santos, F.; Cobo, A.; Álvarez, J.; Rodríguez, J.; Fernández, P. (Oviedo)

Technical Seminars

304 **Isolated Juxtaglomerular Apparatus as a Tool for Exploring Glomerular Hemodynamics: Application of Microperfusion Techniques**
Arima, S.; Ito, S. (Sendai)

312 **Sensitive Analysis of Apoptosis Using Confocal Laser Scan Microscopy**
Oite, T. (Niigata)

No. 6

Renal Ion Channels: Function and Regulation
Guest Editor: Rajesh V. Thakker, Oxford

319 **Preface**
Thakker, R.V. (Oxford)

320 **Disorders of the Epithelial Na^+ Channel in Liddle's Syndrome and Autosomal Recessive Pseudohypoaldosteronism Type 1**
Oh, Y.S.; Warnock, D.G. (Birmingham, Ala.)

326 **Defective Processing and Trafficking of Water Channels in Nephrogenic Diabetes insipidus**
Kamsteeg, E.J.; Deen, P.M.T.; van Os, C.H. (Nijmegen)

332 **Defects in Processing and Trafficking of Cystic Fibrosis Transmembrane Conductance Regulator**
Kunzelmann, K. (St. Lucia, Brisbane); Nitschke, R. (Freiburg)

343 **Mechanisms of Renal Calcium Transport**
Friedman, P.A. (Pittsburgh, Pa.)

351 **Molecular Pathology of Renal Chloride Channels in Dent's Disease and Bartter's Syndrome**
Thakker, R.V. (Oxford)

361 **Severely Impaired Urine-Concentrating Ability in Mice Lacking the CLC-K1 Chloride Channel**
Uchida, S.; Marumo, F. (Tokyo)

366 **Molecular Characteristics of Phosphate Transporters and Their Regulation**
Hernando, N.; Forster, I.C.; Biber, J.; Murer, H. (Zurich)

376 **Author Index Vol. 8, No. 6, 2000**
376 **Subject Index Vol. 8. No. 6, 2000**

377 **Author Index Vol. 8, 2000**
379 **Subject Index Vol. 8, 2000**

Exploring a new field of research

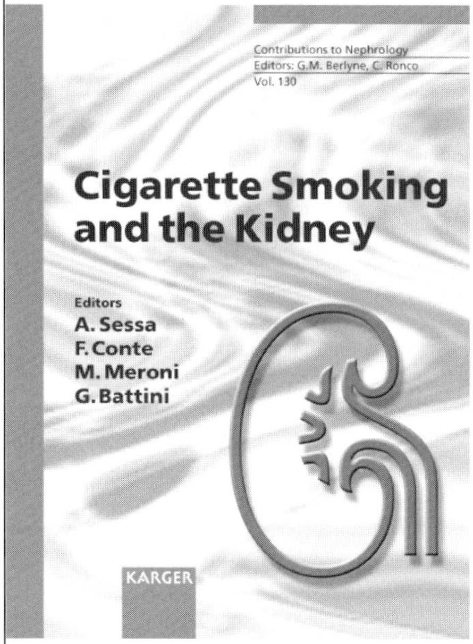

Cigarette Smoking and the Kidney

Editors

Adalberto Sessa
Ferruccio Conte
Mietta Meroni
Graziana Battini

Nephrology; Cardiovascular System; Oncology; Arteriosclerosis, Molecular Biology, Genetics

There is considerable evidence suggesting that chronic cigarette smoking has adverse health effects: it is the major cause of disease and death in developed countries, inducing several kinds of cancer, chronic obstructive lung disease and severe vascular pathologies.
Cigarette smoking is also a risk factor for diffuse atherosclerosis, ischemic heart disease and peripheral vascular lesions, renal microvessels being another potential target. Furthermore, the relationships of chronic cigarette smoking and nitric oxide with endothelial dysfunction and vascular disease are well proven, and the association between eNOS gene polymorphisms and vascular disease and their dependence on chronic tobacco smoking has been stressed. But although there is definite evidence that the risk of progression of acquired or genetic renal disease is higher in chronic cigarette smokers and the negative effect on patient survival in chronic dialytic treatment is well known, these issues are still neglected by clinical nephrologists and the mechanisms involved have not yet been classified.
It is for these reasons that the present volume has been put together: the questions addressed will be of particular interest to renal researchers working on an experimental basis as well as for oncologists and toxicologists.

Contents

Introduction: *Sessa, A.*
The Cultural History of Smoking: *Ritz, E.; Orth, S.R.*
Smoking and Health, with a Note on Kidney Cancer: *Tavani, A.; La Vecchia, C.*
Effects of Cigarette Smoking on Systemic Haemodynamics: *Grassi, G.; Seravalle, G.; Magnoni, M.; Mancia, G.*
Acute Effects of Cigarette Smoking on Renal Hemodynamics: *Ritz, E.; Benck, U.; Orth, S.R.*
Effects of Chronic Cigarette Smoking on the Kidney: *Gambaro, G.; Bertaglia, G.; Citron, L.; Del Prete, D.*
Effect of Cigarette Smoking on Renal Function and Vascular Endothelium: *Remuzzi, G.*
Cigarette Smoking, DNA Variants in Endothelial Nitric Oxide Synthase Gene and Vascular Disease: *Wang, X.L.*
Ischemic Renal Disease: Impact of Cardiovascular Risk Factors and Smoking: *Baggio, B.*
Cigarette Smoking and Early Stage Primary Renal Disease: *Schiff, H.; Lang, S.M.; Bergner, A.*
Cigarette Smoking and Diabetic Nephropathy: *Krimholtz, M.; Thomas, S.; Vibreti, G.C.*
Cigarette Smoking and Other Inhaled Toxins in Anti-GBM Disease: *Hérody, M.; Duvic, C.; Noël, L.-H.; Nédélec, G.; Grünfeld, J.-P.*
Cigarette Smoking and Pauci-Immune Extracapillary Glomerulonephritis with ANCA-Associated Idiopathic Systemic Vasculitis. A Retrospective Study: *Sessa, A.; Meroni, M.; Battini, G.; Vaccari, M.; Giordano, F.; Torri Tarreli, L.*
Smoking as a Risk Factor for End-Stage Renal Failure in Patients with Primary Renal Disease: *Orth, S.R.; Odoni, G.; Ogata, H.; Ritz, E.*
The Effect of Smoking on Urinary NO_x and cGMP Excretion in IgA Nephropathy and in Health: *Kovács, T.; Vas, T.; Wagner, L.; Schmelczer, M.; Kocsis, B.; Nagy, J.*
Lipoperoxidation in Hemodialysis Patients: Upregulation of 5-Lipoxygenase Activity and Expression in Peripheral Blood Mononuclear Cells: *Taccone-Gallucci, T.; Maccarrone, M.; Manca di Villahermosa, S.; Meloni, C.; Finazzi Agrò, A.*
Tobacco Smoking and Renal-Cell Carcinoma. An Overview: *Bulgheroni, P.; Costa, D.; Sbarbati, M.; Bulgheroni, A.*

www.karger.com/bookseries/conep

KI 00646

Contributions to Nephrology, Vol. 130
Series Editors: Berlyne, G.M. (Brooklyn, N.Y./Beersheva); Ronco, C. (Vicenza)
ISSN 0302–5144

Cigarette Smoking and the Kidney
Editors: Sessa, A.; Conte, F.; Meroni, M.; Battini, G. (Vimercate)
VIII + 140 p., 13 fig., 23 tab., hard cover, 2000
CHF 169.– / DEM 219.– / USD 147.00
Prices subject to change
DEM price for Germany, USD price for USA only
ISBN 3–8055–7064–3
Distributed in Japan exclusively by Igaku Shoin Ltd., Tokyo

KARGER

Order Form

Please send: ____ copy/ies

Postage and handling free with prepayment

Payment:
☐ Check enclosed ☐ Please bill me

Please charge to my credit card
☐ American Express ☐ Diners ☐ Eurocard
☐ MasterCard ☐ Visa

Card No.: ____________

Exp. date: ____________

Orders may be placed with any bookshop, subscription agency, directly with the publisher or through a Karger distributor.

Fax to: +41 61 306 12 34

S. Karger AG, P.O. Box, CH–4009 Basel (Switzerland)
E-Mail orders@karger.ch, **www.karger.com**

Name/Address: ____________

Date: ____________

Signature: ____________